Probability: The Logic of Science

Probability: The Logic of Science

Nash Stokes

STATES
ACADEMIC PRESS
www.statesacademicpress.com

Probability: The Logic of Science
Nash Stokes
ISBN: 978-1-63989-445-1 (Hardback)

STATES
ACADEMIC PRESS

Published by States Academic Press,
109 South 5th Street,
Brooklyn, NY 11249, USA

Cataloging-in-Publication Data

Probability : the logic of science / Nash Stokes.
p. cm.
Includes bibliographical references and index.
ISBN 978-1-63989-445-1
1. Probabilities. 2. Mathematical statistics. I. Stokes, Nash.
QA273 .I58 2022
519.2--dc23

For more information regarding States Academic Press and its products, please visit the publisher's website www.statesacademicpress.com

TABLE OF CONTENTS

PREFACE

This book has been written, keeping in view that students want more practical information. Thus, my aim has been to make it as comprehensive as possible for the readers. I would like to extend my thanks to my family and co-workers for their knowledge, support and encouragement all along.

Probability is the numeric description of how likely an event is to occur and how likely it is that the proposition is true. It occurs between zero and one, where zero shows impossibility and one indicates certainty. Practically, there are two categories of probability interpretation. These are objectivists and subjectivists. Objectivists allocate numbers to define some physical or objective state of affairs whereas subjectivists allocate number per subjective probability. The field finds its applications in other areas such as computer science, philosophy, mathematics, artificial intelligence/machine learning, finance, statistics and game theory. It is also used in risk assessment and modelling, biology, ecology and financial management. The topics included in this book on probability are of utmost significance and bound to provide incredible insights to readers. It is compiled in such a manner, that it will provide an in-depth knowledge about the theory and practice of this discipline. In this book, constant effort has been made to make the understanding of the difficult concepts of this field as easy and informative as possible, for the readers.

A brief description of the chapters is provided below for further understanding:

Chapter - What is Probability?

Probability refers to the possibility of an event to occur and it lies between 0 and 1. 0 means impossibility and 1 means certainty. Subjective probability, Bayesian probability and frequentist probability are studied under its domain. This chapter has been carefully written to provide an easy understanding of probability.

Chapter - Probability Theory

The branch of mathematics that is associated with probability is termed as probability theory. Probability space, elementary event, complementary event, independent event, mutual exclusivity, Boole's inequality, etc. are a few of its aspects. This is an introductory chapter which will briefly introduce about probability theory.

Chapter - Conditional Probability

A type of probability in which the probability of an event occurs with a relationship to one or more events is known as conditional probability. It includes conditional probability table, conditional expectation, Lewis's triviality result, conditional variance, etc. This chapter delves into conditional probability for an in-depth understanding of the subject.

Chapter - Probability Distributions

A mathematical and statistical function that describes all the possible values and likelihood that a random variable can have is called probability distribution. Cumulative probability distribution, discrete probability distribution, continuous probability distribution, etc. are studied under its domain. This chapter discusses the subject of probability distribution in detail.

Chapter - Stochastic Processes

A collection of random variables that are defined on a common probability space is called a stochastic process. It involves Poisson point process, gamma process, branching process, Galton-Watson process, Markov process, etc. This chapter sheds light on different stochastic processes for a thorough understanding of the subject.

Nash Stokes

What is Probability? 1

• Interpretations of Probability

Probability refers to the possibility of an event to occur and it lies between 0 and 1. 0 means impossibility and 1 means certainty. Subjective probability, Bayesian probability and frequentist probability are studied under its domain. This chapter has been carefully written to provide an easy understanding of probability.

Probability means possibility. It is a branch of mathematics that deals with the occurrence of a random event. The value is expressed between zero and one. Probability has been introduced in Maths to predict how likely events are to happen. The meaning of probability is basically the extent to which something is likely to happen. This is the basic probability theory which is also used in the probability distribution, where you will learn the possibility of outcomes for a random experiment. To find the probability of a single event to occur, first we should know the total number of possible outcomes.

Probability is a measure of the likelihood of an event to occur. Many events cannot be predicted with total certainty. We can predict only the chance of an event to occur i.e. how likely they are to happen, using it. Probability can range in between 0 to 1, where 0 means the event to be an impossible one and 1 indicates a certain event. Probability for class 10 is an important topic for the students which explains all the basic concepts of this topic. The probability of all the events in a sample space sums up to 1. For example, when we toss a coin, either we get Head or Tail, only two possible outcomes are possible (H, T). But if we toss two coins in the air, there could be three possibilities of events to occur, such as both the coins show heads or both shows tails or one shows heads and one tail, i.e.(H,H), (H,T),(T,T).

Formula for Probability

The probability formula is defined as the possibility of an event to happen is equal to the ratio of the number of favourable outcomes and the total number of outcomes.

> Probability of event to happen P(E) = Number of favourable outcomes/Total Number of outcomes

Example: There are 6 pillows in a bed, 3 are red, 2 are yellow and 1 is blue. What is the probability of picking a yellow pillow?

Solution: The probability is equal to the number of yellow pillows in the bed divided by the total number of pillows, i.e. 2/6 = 1/3.

Example: There is a container full of coloured bottles, red, blue, green and orange. Some of the bottles are picked out and displaced. Sumit did this 1000 times and got the following results:

- No. of blue bottles picked out: 300
- No. of red bottles : 200
- No. of green bottles : 450
- No. of orange bottles : 50.

a) What is the probability that Sumit will pick a green bottle?

Solution: For every 1000 bottles picked out, 450 are green.

Therefore, P(green) = 450/1000 = 0.45

b) If there are 100 bottles in the container, how many of them are likely to be green?

Solution: The experiment implies that 450 out of 1000 bottles are green.

Therefore, out of 100 bottles, 45 are green.

Probability Tree

The tree diagram helps to organize and visualize the different possible outcomes. Branches and ends of the tree are two main positions. Probability of each branch is written on the branch, whereas the ends are containing the final outcome. Tree diagram used to figure out when to multiply and when to add.

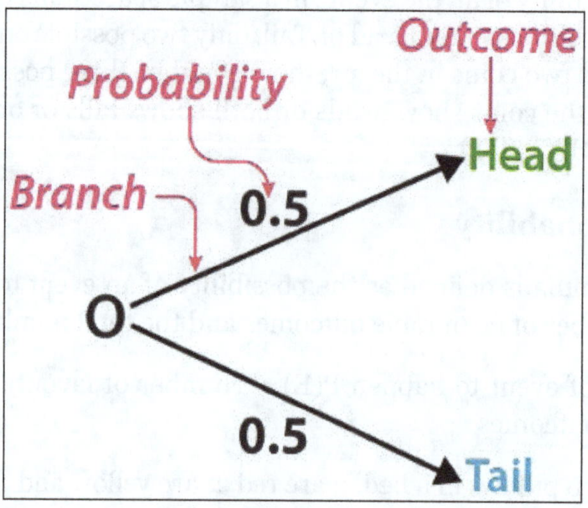

Types of Probability

There are three major types of probabilities:

- Theoretical Probability: It is based on the possible chances of something to happen. The theoretical probability is mainly based on the reasoning behind probability. For example, if a coin is tossed, the theoretical probability of getting head will be ½.
- Experimental Probability: It is based on the basis of the observations of an experiment. The experimental probability can be calculated based on the number of possible outcomes by the total number of trials. For example, if a coin is tossed 10 times and heads is recorded 6 times then, the experimental probability for heads is 6/10 or, 3/5.
- Axiomatic Probability: In axiomatic probability, a set of rules or axioms are set which applies to all types. These axioms are set by Kolmogorov and are known as Kolmogorov's three axioms. With the axiomatic approach to probability, the chances of occurrence or non-occurrence of the events can be quantified. Conditional Probability is the likelihood of an event or outcome occurring based on the occurrence of a previous event or outcome.

Probability of an Event

Assume an event E can occur in r ways out of a sum of n probable or possible equally likely ways. Then the probability of happening of the event or its success is expressed as:

$$P(E) = r/n$$

The probability that the event will not occur or known as its failure is expressed as:

$$P(E') = n - r/n = 1 - r/n$$

E' represents that the event will not occur.

Therefore, now we can say;

$$P(E) + P(E') = 1$$

This means that the total of all the probabilities in any random test or experiment is equal to 1.

Equally Likely Events

When the events have the same theoretical probability of happening, then they are called equally likely events. The results of a sample space are called equally likely if all of them have the same probability of occurring. For example, if you throw a die, then the probability of getting 1 is 1/6. Similarly, the probability of getting all the numbers from 2, 3, 4, 5 and 6, one at a time is 1/6. Hence, the following are some examples of equally likely events when throwing a die:

- Getting 3 and 5 on throwing a die.
- Getting an even number and an odd number on a die.
- Getting 1, 2 or 3 on rolling a die.

These are equally likely events, since the probabilities of each event are equal.

Complementary Events

The possibility that there will be only two outcomes which states that an event will occur or not. Like a person will come or not come to your house, getting a job or not getting a job, etc. are examples of complementary events. Basically, the complement of an event occurring in the exact opposite that the probability of it is not occurring. Some more examples are:

- It will rain or not rain today.
- The student will pass the exam or not pass.
- You win the lottery or you don't.

Probability Density Function

The Probability Density Function(PDF) is the probability function which is represented for the density of a continuous random variable lying between a certain range of values. Probability Density Function explains the normal distribution and how mean and deviation exists. The standard normal distribution is used to create a database or statistics, which are often used in science to represent the real-valued variables, whose distribution are not known.

Interpretations of Probability

Kolmogorov's Probability Calculus

Probability theory was a relative latecomer in intellectual history. To be sure, proto-probabilistic ideas concerning evidence and inference date back to antiquity. However, probability's mathematical treatment had to wait until the Fermat-Pascal correspondence, and their analysis of games of chance in 17th century France. Its axiomatization had to wait still longer, in Kolmogorov's classic Foundations of the Theory of Probability. Roughly, probabilities lie between 0 and 1 inclusive, and they are additive. More formally, let Ω be a non-empty set ('the universal set'). A field (or algebra) on Ω is a set **F** of subsets of Ω that has Ω as a member, and that is closed under complementation (with respect to Ω) and union. Let P be a function from **F** to the real numbers obeying:

- (Non-negativity) $P(A) \geq 0$, for all $A \in \mathbf{F}$.
- (Normalization) $P(\Omega) = 1$.
- (Finite additivity $P(A \cup B) = P(A) + P(B)$ for all $A, B \in \mathbf{F}$ such that $A \cap B = \varnothing$.

Call PP a probability function, and $(\Omega, \mathbf{F}, P)$ a probability space. This is Kolmogorov's "elementary theory of probability".

The assumption that P is defined on a field guarantees that these axioms are non-vacuously instantiated, as are the various theorems that follow from them. The non-negativity and normalization axioms are largely matters of convention, although it is non-trivial that probability functions take at least the two values 0 and 1, and that they have a maximal value (unlike various other measures, such as length, volume, and so on, which are unbounded).

We may now apply the theory to various familiar cases. For example, we may represent the results of tossing a single die once by the set $\Omega = \{1,2,3,4,5,6\}$, and we could let **F** be the set of all subsets of Ω. Under the natural assignment of probabilities to members of **F**, we obtain such welcome results as the following:

$$
\begin{aligned}
P(\{1\}) &= \frac{1}{6}, \\
P(\text{even}) &= P(\{2\} \cup \{4\} \cup \{6\}) \\
&= \frac{3}{6}, \\
P(\text{odd or less than } 4) &= P(\text{odd}) + P(\text{less than } 4) - P(\text{odd} \cap \text{less than } 4) \\
&= \frac{1}{2} + \frac{1}{2} - \frac{2}{6} \\
&= \frac{4}{6},
\end{aligned}
$$

and so on.

We could instead attach probabilities to members of a collection **S** of sentences of a formal language, closed under (countable) truth-functional combinations, with the following counterpart axiomatization:

- $P(A) \geq 0$ for all $A \in \mathbf{S}$.

- If T is a logical truth (in classical logic), then $P(T)=1$.
- $P(A \vee B)=P(A)+P(B)$ for all $A \in \mathbf{S}$ and $B \in \mathbf{S}$ such that A and B are logically incompatible.

The bearers of probabilities are sometimes also called “events”, “outcomes”, or “propositions”, but the underlying formalism remains the same. More attention has been given to interpreting ‘P’ than to interpreting its bearers; we will be concerned with the former.

Now let us strengthen our closure assumptions regarding **F**, requiring it to be closed under complementation and countableunion; it is then called a sigma field (or sigma algebra) on Ω. It is controversial whether we should strengthen finite additivity, as Kolmogorov does:

Countable Additivity

If $A_1, A_2, A_3 \ldots$ is a countably infinite sequence of (pairwise) disjoint sets, each of which is an element of **F**, then:

$$P\left(\bigcup_{n=1}^{\infty} A_n\right)=\sum_{n=1}^{\infty} P(A_n)$$

Kolmogorov comments that infinite probability spaces are idealized models of real random processes, and that he limits himself arbitrarily to only those models that satisfy countable additivity. This axiom is the cornerstone of the assimilation of probability theory to measure theory.

The conditional probability of A given B is then given by the ratio of unconditional probabilities:

$$P(A \mid B)=\frac{P(A \cap B)}{P(B)}, \text{provided } P(B)>0.$$

This is often taken to be the definition of conditional probability, although it should be emphasized that this is a technical usage of the term that may not align perfectly with a pretheoretical concept that we might have. We recognize it in locutions such as “the probability that the die lands 1, given that it lands odd, is 1/3”, or “the probability that it will rain tomorrow, given that there are dark clouds in the sky tomorrow morning, is high”. It is the concept of the probability of something given or in the light of some piece of evidence or information. Indeed, some authors take conditional probability to be the primitive notion, and axiomatize it directly.

There are other formalizations that give up normalization; that give up countable additivity and even additivity; that allow probabilities to take infinitesimal values (positive, but smaller than every positive real number) that allow probabilities to be imprecise interval-valued, or more generally represented with sets of precise probability functions; and that treat probabilities comparatively rather than quantitatively. For now, however, when we speak of 'the probability calculus', we will mean Kolmogorov's approach, as is standard. Given certain probabilities as inputs, the axioms and theorems allow us to compute various further probabilities. However, apart from the assignment of 1 to the universal set and 0 to the empty set, they are silent regarding the initial assignment of probabilities. For guidance with that, we need to turn to the interpretations of probability. First, however, let us list some criteria of adequacy for such interpretations.

Criteria of Adequacy for the Interpretations of Probability

What criteria are appropriate for assessing the cogency of a proposed interpretation of probability? Of course, an interpretation should be precise, unambiguous, non-circular, and use well-understood primitives. But those are really prescriptions for good philosophizing generally; what do we want from our interpretations of probability, specifically? We begin by following Salmon, although we will raise some questions about his criteria, and propose some others. He writes:

- Admissibility: We say that an interpretation of a formal system is admissible if the meanings assigned to the primitive terms in the interpretation transform the formal axioms, and consequently all the theorems, into true statements. A fundamental requirement for probability concepts is to satisfy the mathematical relations specified by the calculus of probability.

- Ascertainability: This criterion requires that there be some method by which, in principle at least, we can ascertain values of probabilities. It merely expresses the fact that a concept of probability will be useless if it is impossible in principle to find out what the probabilities are.

- Applicability: The force of this criterion is best expressed in Bishop Butler's famous aphorism, "Probability is the very guide of life."

It might seem that the criterion of admissibility goes without saying. The word 'interpretation' is often used in such a way that 'admissible interpretation' is a pleonasm. Yet it turns out that the criterion is non-trivial, and indeed if taken seriously would rule out several of the leading interpretations of probability. Some of them fail to satisfy countable additivity; for others (certain propensity interpretations) the status of at least some of the axioms is unclear. Nevertheless, we regard them as genuine candidates. It should be remembered, moreover, that Kolmogorov's is just one of many possible axiomatizations, and there is not universal agreement on which is 'best' (whatever that might mean). Indeed, Salmon's preferred axiomatization differs from Kolmogorov's. Thus, there is no such thing as admissibility tout court, but rather admissibility with respect to this or that axiomatization. It would be unfortunate if, perhaps out of an overdeveloped regard for history, one felt obliged to reject any interpretation that did not obey the letter of Kolmogorov's laws and that was thus 'inadmissible'. In any case, if we found an inadmissible interpretation that did a wonderful job of meeting the criteria of ascertainability and applicability, then we should surely embrace it.

Most of the work will be done by the applicability criterion. We must say more (as Salmon indeed does) about what *sort* of a guide to life probability is supposed to be. Mass, length, area and volume are all useful concepts, and they are 'guides to life' in various ways (think how critical distance judgments can be to survival); moreover, they are admissible and ascertainable, so presumably it is the applicability criterion that will rule them out. Perhaps it is best to think of applicability as a cluster of criteria, each of which is supposed to capture something of probability's distinctive conceptual roles; moreover, we should not require that all of them be met by a given interpretation. They include:

- Non-triviality: An interpretation should make non-extreme probabilities at least a conceptual possibility. For example, suppose that we interpret 'P' as the truth function: it assigns the value 1 to all true sentences, and 0 to all false sentences. Then trivially, all the axioms come out true, so this interpretation is admissible. We would hardly count it as an adequate interpretation of probability, however, and so we need to exclude it. It is essential to probability that, at least in principle, it can take intermediate values.

- Applicability to frequencies: An interpretation should render perspicuous the relationship between probabilities and (long-run) frequencies. Among other things, it should make clear why, by and large, more probable events occur more frequently than less probable events.

- Applicability to rational beliefs: an interpretation should clarify the role that probabilities play in constraining the degrees of belief, or credences, of rational agents. Among other things, knowing that one event is more probable than another, a rational agent will be more confident about the occurrence of the former event.

- Applicability to rational decisions: An interpretation should make clear how probabilities figure in rational decision-making. This seems especially apposite for a 'guide to life'.

- Applicability to ampliative inferences: An interpretation will score bonus points if it illuminates the distinction between 'good' and 'bad' ampliative inferences, while explicating why both fall short of deductive inferences.

- Applicability to science: An interpretation should illuminate paradigmatic uses of probability in science (for example, in quantum mechanics and statistical mechanics).

Broadly speaking, there are arguably three main concepts of probability:

- An epistemological concept, which is meant to measure objective evidential support relations. For example, "in light of the relevant seismological and geological data, California will probably experience a major earthquake this decade".

- The concept of an agent's degree of confidence, a graded belief. For example, "I am not sure that it will rain in Canberra this week, but it probably will."
- A physical concept that applies to various systems in the world, independently of what anyone thinks. For example, "a particular radium atom will probably decay within 10,000 years".

Classical Probability

The classical interpretation owes its name to its early and august pedigree. It was championed by de Moivre and Laplace, and inchoate versions of it may be found in the works of Pascal, Bernoulli, Huygens, and Leibniz. It assigns probabilities in the absence of any evidence, or in the presence of symmetrically balanced evidence. The guiding idea is that in such circumstances, probability is shared equally among all the possible outcomes, so that the classical probability of an event is simply the fraction of the total number of possibilities in which the event occurs. It seems especially well suited to those games of chance that by their very design create such circumstances — for example, the classical probability of a fair die landing with an even number showing up is 3/6. It is often presupposed (usually tacitly) in textbook probability puzzles.

Here is a classic statement by de Moivre:

> "[I]f we constitute a fraction whereof the numerator be the number of chances whereby an event may happen, and the denominator the number of all the chances whereby it may either happen or fail, that fraction will be a proper designation of the probability of happening."

Laplace gives the best-known but slightly different formulation:

> "The theory of chances consists in reducing all events of the same kind to a certain number of equally possible cases, that is to say, to cases whose existence we are equally uncertain of, and in determining the number of cases favourable to the event whose probability is sought. The ratio of this number to that of all possible cases is the measure of this probability, which is thus only a fraction whose numerator is the number of favourable cases, and whose denominator is the number of all possible cases."

We may ask a number of questions about this formulation. When are events of the same kind? Intuitively, 'heads' and 'tails' are equally likely outcomes of tossing a fair coin; but if their kind is 'ways the coin could land', then 'edge' should presumably be counted alongside them. The "certain number of equally possible cases" and "that of all possible cases" are presumably finite numbers. What, then, of probabilities in infinite spaces? Apparently, irrational-valued probabilities such as $1/\sqrt{2}$ are automatically eliminated, and thus theories such as quantum mechanics that posit them cannot be accommodated.

Who are "we", who "are equally uncertain"? Different people may be equally undecided about different things, which suggests that Laplace is offering a subjectivist interpretation in which probabilities vary from person to person depending on contingent differences in their evidence. Yet he means to characterize the objective probability assignment of a rational agent in an epistemically neutral position with respect to a set of "equally possible" cases. But then the proposal risks sounding empty: for what is it for an agent to *be* "equally uncertain" about a set of cases, other than assigning them equal probability?

This brings us to one of the key objections to Laplace's account. The notion of "equally possible" cases faces the charge of either being a category mistake (for 'possibility' does not come in degrees), or circular (for what is meant is really 'equally probable'). The notion is finessed by the so-called 'principle of indifference', a coinage due to Keynes (although he was no friend of the principle): "if there is no known reason for predicating of our subject one rather than another of several alternatives, then relatively to such knowledge the assertions of each of these alternatives have an equal probability". (The 'principle of equal probability' would be a better name). Thus, it might be claimed, there is no circularity in the classical interpretation after all. However, this move may only postpone the problem, for there is still a threat of circularity, albeit at a lower level. We have two cases here: outcomes for which we have no evidence ("reason") at all, and outcomes for which we have symmetrically balanced evidence. There is no circularity in the first case unless the notion of 'evidence' is itself probabilistic; but artificial examples aside, it is doubtful that the case ever arises. For example, we have a considerable fund of evidence on coin tossing from the results of our own experiments, the testimony of others, our knowledge of some of the relevant physics, and so on. In the second case, the threat of circularity is more apparent, for it seems that some sort of weighing of the evidence in favor of each outcome is required, and this seems to require a reference to probability. Indeed, the most obvious characterization of symmetrically balanced evidence is in terms of equality of conditional probabilities: given evidence E and possible outcomes, $O_1, O_2, \dots, O_n$, the evidence is symmetrically balanced iff $P(O_1|E) = P(O_2|E) = \dots = P(O_n|E)$. Then it seems that probabilities reside at the base of the interpretation after all. Still, it would be an achievement if all probabilities could be reduced to cases of equal probability.

Laplace's classical theory is restricted to finite spaces, one for which there are only finitely many possible outcomes. When the spaces are countably infinite, the spirit of the classical theory may be upheld by appealing to the information-theoretic principle of maximum entropy, a generalization of the principle of indifference championed by Jaynes. Entropy is a measure of the lack of 'informativeness' of a probability function. The more concentrated is the function, the less is its entropy; the more diffuse it is, the greater is its entropy. For a discrete assignment of probabilities $P = (p_1, p_2, \ldots)$, the entropy of P is defined as:

$$-\sum_i p_i \log p_i$$

The principle of maximum entropy enjoins us to select from the family of all probability functions consistent with our background knowledge the function that maximizes this quantity. In the special case of choosing the most uninformative probability function over a finite set of possible outcomes, this is just the familiar 'flat' classical assignment. Things get more complicated in the infinite case, since there cannot be a flat assignment over denumerably many outcomes, on pain of violating the standard probability calculus (with countable additivity). Rather, the best we can have are sequences of progressively flatter assignments, none of which is truly flat. We must then impose some further constraint that narrows the field to a smaller family in which there is an assignment of maximum entropy. This constraint has to be imposed from outside as background knowledge, but there is no general theory of which external constraint should be applied when.

Let us turn now to uncountably infinite spaces. It is easy — all too easy — to assign equal probabilities to the points in such a space: each gets probability 0. Non-trivial probabilities arise when uncountably many of the points are clumped together in larger sets. If there are finitely many clumps, Laplace's classical theory may be appealed to again: if the evidence bears symmetrically on these clumps, each gets the same share of probability.

Bertrand's paradoxes arise in uncountable spaces and turn on alternative parametrizations of a given problem that are non-linearly related to each other. Some presentations are needlessly arcane; length and area suffice to make the point. The following example nicely illustrates how Bertrand-style paradoxes work. A factory produces cubes with side-length between 0 and 1 foot; what is the probability that a randomly chosen cube has side-length between 0 and 1/2 a foot? The classical intepretation's answer is apparently 1/2, as we imagine a process of production that is uniformly distributed over side-length. But the question could have been given an equivalent restatement: A factory produces cubes with face-area between 0 and 1 square-feet; what is the probability that a randomly chosen cube has face-area between 0 and 1/4 square-feet? Now the answer is apparently 1/4, as we imagine a process of production that is uniformly distributed over face-area. This is already disastrous, as we cannot allow the same event to have two different probabilities (especially if this interpretation is to be admissible). But there is worse to come, for the problem could have been restated equivalently again: A factory produces cubes with volume between 0 and 1 cubic feet; what is the probability that a randomly chosen cube has volume between 0 and 1/8 cubic-feet? Now the answer is apparently 1/8, as we imagine a process of production that is uniformly distributed over volume. And so on for all of the infinitely many equivalent reformulations of the problem (in terms of the fourth, fifth, ... power of the length, and indeed in terms of every non-zero real-valued exponent of the length). What, then, is *the* probability of the event in question?

The paradox arises because the principle of indifference can be used in incompatible ways. We have no evidence that favors the side-length lying in the interval [0, 1/2] over its lying in [1/2, 1], or vice versa, so the principle requires us to give probability 1/2 to each. Unfortunately, we also have no evidence that favors the face-area lying in any of the four intervals [0, 1/4], [1/4, 1/2], [1/2, 3/4], and [3/4, 1] over any of the others, so we must give probability 1/4 to each. The event 'the side-length lies in [0, 1/2]', receives a different probability when merely redescribed. And so it goes, for all the other reformulations of the problem. We cannot meet any pair of these constraints simultaneously, let alone all of them.

Jaynes attempts to save the principle of indifference and to extend the principle of maximum entropy to the continuous case, with his invariance condition: in two problems where we have the same knowledge, we should assign the same probabilities. He regards this as a consistency requirement. For any problem, we have a group of admissible transformations, those that change the problem into an equivalent form. Various details are left unspecified in the problem; equivalent formulations of it fill in the details in different ways. Jaynes' invariance condition bids us to assign equal probabilities to equivalent propositions, reformulations of one another that are arrived at by such admissible transformations of our problem. Any probability assignment that meets this condition is called an invariant assignment. Ideally, our problem will have a unique invariant assignment. To be sure, things will not always be ideal; but sometimes they are, in which case this is surely progress on Bertrand-style problems.

For many garden-variety problems such technical machinery will not be needed. Suppose we tell you that a prize is behind one of three doors, and you get to choose a door. This seems to be a paradigm case in which the principle of indifference works well: the probability that you choose the right door is 1/3. It seems implausible that we should worry about some reparametrization of the problem that would yield a different answer. To be sure, Bertrand-style problems caution us that there are limits to the principle of indifference. But arguably we must just be careful not to overstate its applicability. How does the classical theory of probability fare with respect to our criteria of adequacy? Let us begin with admissibility. (Laplacean) classical probabilities obey non-negativity and normalization, but they are only finitely additive. So they do not obey the full Kolmogorov probability calculus, but they provide an interpretation of the elementary theory.

Classical probabilities are ascertainable, assuming that the space of possibilities can be determined in principle. They bear a relationship to the credences of rational agents; the circularity concern, as we saw above, is that the relationship is vacuous, and that rather than constraining the credences of a rational agent in an epistemically neutral position, they merely record them.

Without supplementation, the classical theory makes no contact with frequency information. However the coin happens to land in a sequence of trials, the possible outcomes remain the same. Indeed, even if we have strong empirical evidence that the coin is biased towards heads with probability, say, 0.6, it is hard to see how the unadorned classical theory can accommodate this fact – for what now are the ten possibilities, six of which are favorable to heads? Laplace does supplement the theory with his Rule of Succession: "Thus we find that an event having occurred successively any number of times, the probability that it will happen again the next time is equal to this number increased by unity divided by the same number, increased by two units." That is:

$$\Pr(\text{success on N} + 1^{\text{st}} \text{ trial} \mid \text{N consec. succeses}) = \frac{N+1}{N+2}$$

Science apparently invokes at various points probabilities that look classical. Bose-Einstein statistics, Fermi-Dirac statistics, and Maxwell-Boltzmann statistics each arise by considering the ways in which particles can be assigned to states, and then applying the principle of indifference to different subdivisions of the set of alternatives, Bertrand-style. The trouble is that Bose-Einstein statistics apply to some particles (e.g. photons) and not to others, Fermi-Dirac statistics apply to different particles (e.g. electrons), and Maxwell-Boltzmann statistics do not apply to any known particles. None of this can be determined a priori, as the classical interpretation would have it. Moreover, the classical theory purports to yield probability assignments in the face of ignorance. But as Fine writes:

> "If we are truly ignorant about a set of alternatives, then we are also ignorant about combinations of alternatives and about subdivisions of alternatives. However, the principle of indifference when applied to alternatives, or their combinations, or their subdivisions, yields different probability assignments."

This brings us to one of the chief points of controversy regarding the classical interpretation. Critics accuse the principle of indifference of extracting information from ignorance. Proponents reply that it rather codifies the way in which such ignorance should be epistemically managed – for anything other than an equal assignment of probabilities would represent the possession of some knowledge. Critics counter-reply that in a state of complete ignorance, it is better to assign imprecise probabilities (perhaps ranging over the entire [0, 1] interval), or to eschew the assignment of probabilities altogether.

The Logical Interpretation

Logical theories of probability retain the classical interpretation's idea that probabilities can be determined a priori by an examination of the space of possibilities. However, they generalize it in two important ways: the possibilities may be assigned unequal weights, and probabilities can be computed whatever the evidence may be, symmetrically balanced or not. Indeed, the logical interpretation, in its various guises, seeks to encapsulate in full generality the degree of support or confirmation that a piece of evidence e confers upon a given hypothesis h, which we may write as c(h,e). In doing so, it can be regarded also as generalizing deductive logic and its notion of implication, to a complete theory of inference equipped with the notion of 'degree of implication' that relates e to h. It is often called the theory of 'inductive logic', although this is a misnomer: there is no requirement that e be in any sense 'inductive' evidence for h. 'Non-deductive logic' would be a better name, but this overlooks the fact that deductive logic's relations of implication and incompatibility are also accommodated as extreme cases in which the confirmation function takes the values 1 and 0 respectively. In any case, it is significant that the logical interpretation provides a framework for induction.

Early proponents of logical probability include Johnson, Keynes, and Jeffreys. However, by far the most systematic study of logical probability was by Carnap. His formulation of logical probability begins with the construction of a formal language. In he considers a class of very simple languages consisting of a finite number of logically independent monadic predicates (naming properties) applied to countably many individual constants (naming individuals) or variables, and the usual logical connectives. The strongest (consistent) statements that can be made in a given language describe all of the individuals in as much detail as the expressive power of the language allows. They are conjunctions of complete descriptions of each individual, each description itself a conjunction containing exactly one occurrence (negated or unnegated) of each predicate of the language. Call these strongest statements state descriptions.

Any probability measure m(–) over the state descriptions automatically extends to a measure over all sentences, since each sentence is equivalent to a disjunction of state descriptions; m in turn induces a confirmation function c(–,–):

$$c(h,e) = \frac{m(h \,\&\, e)}{m(e)}$$

There are infinitely many candidates for m, and hence c, even for very simple languages. Carnap argues for his favored measure "m*" by insisting that the only thing that significantly distinguishes individuals from one another is some qualitative difference, not just a difference in labeling. Call astructure description a maximal set of state descriptions, each of which can be obtained from another by some permutation of the individual names. m* assigns each structure description equal measure, which in turn is divided equally among their constituent state descriptions. It gives greater weight to homogenous state descriptions than to heterogeneous ones, thus 'rewarding' uniformity among the individuals in accordance with putatively reasonable inductive practice. The induced c* allows inductive learning from experience.

Consider, for example, a language that has three names, a, b and c, for individuals, and one predicate F. For this language, the state descriptions are:

1.	Fa	&	Fb	&	Fc
2.	¬Fa	&	Fb	&	Fc
3.	Fa	&	¬Fb	&	Fc
4.	Fa	&	Fb	&	¬Fc
5.	¬Fa	&	¬Fb	&	Fc
6.	¬Fa	&	Fb	&	¬Fc
7.	Fa	&	¬Fb	&	¬Fc
8.	¬Fa	&	¬Fb	&	¬Fc

There are four structure descriptions:

{1}, "Everything is F";
{2,3,4}, "Two Fs, one ¬F";
{5,6,7}, "One F, two ¬Fs"; and
{8}, "Everything is ¬F";

The measure m* assigns numbers to the state descriptions as follows: first, every structure description is assigned an equal weight, 1/4; then, each state description belonging to a given structure description is assigned an equal part of the weight assigned to the structure description:

State description	Structure Description	Weight	m*
1.Fa.Fb.Fc	I. Everything is F	1/4	1/4
2.¬Fa.Fb.Fc			1/12
3.Fa.¬Fb.Fc	II. Two Fs, one ¬F	1/4	1/12
4.Fa.Fb.¬Fc			1/12
5.¬Fa.¬Fb.Fc			1/12
6.¬Fa.Fb.¬Fc	III. One F, two ¬Fs	1/4	1/12
7.Fa.¬Fb.¬Fc			1/12
8.¬Fa.¬Fb.¬Fc	IV. Everything is ¬F	1/4	1/4

Notice that m* gives greater weight to the homogenous state descriptions 1 and 8 than to the heterogeneous ones. This will manifest itself in the inductive support that hypotheses can gain from appropriate evidence statements. Consider the hypothesis statement $h = Fc$, true in 4 of the 8 state descriptions, with a priori probability $m^*(h) = 1/2$. Suppose we examine individual "a" and find it has property F — call this evidence e.

Intuitively, e is favorable (albeit weak) inductive evidence for h. We have $m^*(h\&e)=1/3$, $m^*(e)=1/2$, and hence:

$$c^*(h,e)=\frac{m^*(h\&e)}{m^*(e)}=\frac{2}{3}.$$

This is greater than the *a priori* probability $m^*(h)=1/2$, so the hypothesis has been confirmed. It can be shown that in general m* yields a degree of confirmation c* that allows learning from experience.

However, that infinitely many confirmation functions, defined by suitable choices of the initial measure, allow learning from experience. We do not have yet a reason to think that c* is the right choice. Carnap claims nevertheless that c* stands out for being simple and natural.

He later generalizes his confirmation function to a continuum of functions c_λ. Define a *family* of predicates to be a set of predicates such that, for each individual, exactly one member of the set applies, and consider first-order languages containing a finite number of families. Carnap focuses on the special case of a language containing only one-place predicates. He lays down a host of axioms concerning the confirmation function cc, including those induced by the probability calculus itself, various axioms of symmetry (for example, that c(h,e) remains unchanged under permutations of individuals, and of predicates of any family), and axioms that guarantee undogmatic inductive learning, and long-run convergence to relative frequencies. They imply that, for a family $\{P_n\}, n=1,\ldots,k\,(k>2)$:

$$c\lambda(\text{individual } s+1 \text{ is } P_j, s_j \text{ of the first } s \text{ individuals are } P_j)=\frac{(s_j+\lambda/k)}{s+\lambda},$$

where λ is a positive real number. The higher the value of λ, the less impact evidence has: induction from what is observed becomes progressively more swamped by a classical-style equal assignment to each of the k possibilities regarding individual s+1.

Firstly, is there a correct setting of λ, or said another way, how 'inductive' should the confirmation function be? The concern here is that any particular setting of λ is arbitrary in a way that compromises Carnap's claim to be offering a logical notion of probability. Also, it turns out that for any such setting, a universal statement in an infinite universe always receives zero confirmation, no matter what the (finite) evidence. Many find this counterintuitive, since laws of nature with infinitely many instances can apparently be confirmed. Earman discusses the prospects for avoiding the unwelcome result.

Significantly, Carnap's various axioms of symmetry are hardly logical truths. Moreover, Fine argues that we cannot impose further symmetry constraints that are seemingly just as plausible as Carnap's, on pain of inconsistency. Goodman taught us: that the future will resemble the past in some respect is trivial; that it will resemble the past in all respects is contradictory. And we may continue: that a probability assignment can be made to respect some symmetry is trivial; that one can be made to respect all symmetries is contradictory. This threatens the whole program of logical probability.

Another Goodmanian lesson is that inductive logic must be sensitive to the meanings of predicates, strongly suggesting that a purely syntactic approach such as Carnap's is doomed. Scott and Krauss use model theory in their formulation of logical probability for richer and more realistic languages than Carnap's. Still, finding a canonical language seems too many to be a pipe dream, at least if we want to analyze the "logical probability" of any argument of real interest — either in science, or in everyday life.

Logical probabilities are admissible. It is easily shown that they satisfy finite additivity, and given that they are defined on finite sets of sentences, the extension to countable additivity is trivial. Given a choice of language, the values of a given confirmation functions are ascertainable; thus, if this language is rich enough for a given application, the relevant probabilities are ascertainable. The whole point of the theory of logical probability is to explicate ampliative inference, although given the apparent arbitrariness in the choice of language and in the setting of λ — thus, in the choice of confirmation function — one may wonder how well it achieves this. The problem of arbitrariness of the confirmation function also hampers the extent to which the logical interpretation can truly illuminate the connection between probabilities and frequencies.

The arbitrariness problem, moreover, stymies any compelling connection between logical probabilities and rational credences. And a further problem remains even after the confirmation function has been chosen: if one's credences are to be based on logical probabilities, they must be relativized to an evidence statement, e. Carnap requires that e be one's *total evidence*—the maximally specific information at one's disposal, the strongest proposition of which one is certain. But perhaps learning does not come in the form of such 'bedrock' propositions, as Jeffrey has argued — maybe it rather involves a shift in one's subjective probabilities across a partition, without any cell of the partition becoming certain. Then it may be that the strongest proposition of which one is certain is expressed by a tautology T — hardly an interesting notion of 'total evidence'.

In connection with the 'applicability to science' criterion, a point due to Lakatos is telling. By Carnap's lights, the degree of confirmation of a hypothesis depends on the language in which the hypothesis is stated and over which the confirmation function is defined. But scientific progress often brings with it a change in scientific language (for example, the addition of new predicates and the deletion of old ones), and such a change will bring with it a change in the corresponding cc-values. Thus, the growth of science may overthrow any particular confirmation theory. There is something of the snake eating its own tail here, since logical probability was supposed to explicate the confirmation of scientific theories.

We have seen that the later Carnap relaxed his earlier aspiration to find a *unique* confirmation function, allowing a continuum of such functions displaying a wide range of inductive cautiousness. Various critics of logical probabilities believe that he did not go far enough — that even his later systems constrain inductive learning beyond what is rationally required. This recalls the classic debate earlier in the 20th century between Keynes, a famous proponent of logical probabilities, and Ramsey, an equally famous opponent. Ramsey was skeptical of there being any non-trivial relations of logical probability: he said that he could not discern them himself, and that others disagree about them.

The Evidential Interpretation

One might insist, however, that there are non-trivial probabilistic *evidential* relations, even if they are not logical. It may not be a matter of *logic* that the sun will probably rise tomorrow, given our evidence, yet there still seems to be an objective sense in which it probably will, given our evidence. In a crime investigation, there may be a fact of the matter of how strongly the available evidence supports the guilt of various suspects. This does not seem to be a matter of logic—nor of physics, nor of what anyone happens to think, nor of how the facts in the actual world turn out. It seems to be a matter, rather, of evidential probabilities.

More generally, Timothy Williamson writes:

> "Given a scientific hypothesis h, we can intelligibly ask: how probable is h on present evidence? We are asking how much the evidence tells for or against the hypothesis. We are not asking what objective physical chance or frequency of truth h has. A proposed law of nature may be quite improbable on present evidence even though its objective chance of truth is 1. That is quite consistent with the obvious point that the evidence bearing on h may include evidence about objective chances or frequencies. Equally, in asking how probable h is on present evidence, we are not asking about anyone's actual degree of belief in h. Present evidence may tell strongly against h, even though everyone is irrationally certain of h."

Williamson identifies one's evidence with what one knows. However, one might adopt other conceptions of evidence, and one might even take evidential probabilities to link any two propositions whatsoever. Williamson maintains that evidential probabilities are not logical—in particular, they are not syntactically definable. He assumes an initial probability distribution P, which "measures something like the intrinsic plausibility of hypotheses prior to investigation". The evidential probability of h on total evidence e is then given by $P(h \mid e)$.

Are evidential probabilities admissible? Williamson says that "P will be assumed to satisfy a standard set of axioms for the probability calculus". So admissibility is built into the very specification of P. Are they ascertainable? He writes:

> "What, then, are probabilities on evidence? We should resist demands for an operational definition; such demands are as damaging in the philosophy of science as they are in science itself. Sometimes the best policy is to go ahead and theorize with a vague but powerful notion. One's original intuitive understanding

becomes refined as a result, although rarely to the point of a definition in precise pretheoretic terms. That policy will be pursued here."

This might be understood as rejecting ascertainability as a criterion of adequacy. However, some authors are skeptical that there are such things as evidential probabilities—e.g. Joyce. He also argues that there is more than one sense in which evidence tells for or against a hypothesis. Bacon allows that there are such things as evidential probabilities, but he argues that various puzzling results follow from Williamson's account of them, in virtue of its identifying evidence with knowledge. Moreover, one may resist demands for an operational definition of evidential probabilities, while seeking some further understanding of them in terms of other theoretical concepts. For example, perhaps P(h | e) is the subjective probability that a perfectly rational agent with evidence e would assign to h? Williamson argues against this proposal; Eder (forthcoming) defends it, and she offers several ways of interpreting evidential probabilities in terms of ideal subjective probabilities. If some such way is tenable, evidential probabilities would presumably enjoy whatever applicability that such subjective probabilities have.

Subjective Probability

Subjectivists, also known as Bayesians or followers of epistemic probability, give the notion of probability a subjective status by regarding it as a measure of the 'degree of belief' of the individual assessing the uncertainty of a particular situation. Epistemic or subjective probability is sometimes called credence, as opposed to the term chance for a propensity probability. Some examples of epistemic probability are to assign a probability to the proposition that a proposed law of physics is true, and to determine how probable it is that a suspect committed a crime, based on the evidence presented.

Gambling odds reflect the average bettor's 'degree of belief' in the outcome.

Gambling odds don't reflect the bookies' belief in a likely winner, so much as the other bettors' belief, because the bettors are actually betting against one another. The odds are set based on how many people have bet on a possible winner, so that even if the high odds players always win, the bookies will always make their percentages anyway.

The use of Bayesian probability raises the philosophical debate as to whether it can contribute valid justifications of belief. Bayesians point to the work of Ramsey and de Finetti as proving that subjective beliefs must follow the laws of probability if they are to be coherent. Evidence casts doubt that humans will have coherent beliefs.

The use of Bayesian probability involves specifying a prior probability. This may be obtained from consideration of whether the required prior probability is greater or lesser than a reference probability associated with an urn model or a thought experiment. The issue is that for a given problem, multiple thought experiments could apply, and choosing one is a matter of judgement: different people may assign different prior probabilities, known as the reference class problem. The "sunrise problem" provides an example.

Bayesian Probability

Bayesian probability theory provides a mathematical framework for preforming inference, or reasoning, using probability. The foundations of Bayesian probability theory were laid down some 200 years ago by people such as Bernoulli, Bayes, and Laplace, but it has been held suspect or controversial by modern statisticians. The last few decades though have seen the occurrence of a "Bayesian revolution," and Bayesian probability theory is now commonly employed (oftentimes with stunning success) in many scientific disciplines, from astrophysics to neuroscience. It is most often used to judge the relative validity of hypotheses in the face of noisy, sparse, or uncertain data, or to adjust the parameters of a specific model.

Bayes' Rule

Bayes' rule really involves nothing more than the manipulation of conditional probabilities. Remember that the joint probability of two events, A & B, can be expressed as,

$$\begin{aligned} P(AB) &= P(A|B)P(B) \\ &= P(B|A)P(A) \end{aligned}$$

In Bayesian probability theory, one of these "events" is the hypothesis, H, and the other is data, D, and we wish to judge the relative truth of the hypothesis given the data. According to Bayes' rule, we do this via the relation,

$$P(H|D) = \frac{P(D|H)P(H)}{P(D)}$$

The term P(D|H) is called the likelihood function and it assesses the probability of the observed data arising from the hypothesis. Usually this is known by the experimenter, as it expresses one's knowledge of how one expects the data to look given that the hypothesis is true. The term P(H) is called the prior, as it reflects one's prior knowledge before the data are considered. The specification of the prior is often the most subjective aspect of Bayesian probability theory, and it is one of the reasons statisticians held Bayesian inference in contempt. But closer examination of traditional statistical methods reveals that they all have their hidden assumptions and tricks built into them. Indeed, one of the advantages of Bayesian probability theory is that one's assumptions are made up front, and any element of subjectivity in the reasoning process is directly exposed. The term P(D) is obtained by integrating (or summing) P(D|H)P(H) over all H, and usually plays the role of an ignorable normalizing constant. Finally, the term P(H|D) is known as the posterior, and as its name suggests, reflects the probability of the hypothesis after consideration of the data.

Another way of looking at Bayes' rule is that it represents learning. That is, the transformation from the prior, P(H), to the posterior, P(H|D), formally reflects what we have learned about the validity of the hypothesis from consideration of the data. Now let's see how all of this is played out in a rather simplified example.

Example: A classic example of where Bayesian inference is employed is in the problem of estimation, where we must guess the value of an underlying parameter from an observation that is corrupted by noise. Let's say we have some quantity in the world, x, and our observation of this quantity, y, is corrupted by additive Gaussian noise, n, with zero mean:

$$y = x + n$$

Our job is to make the best guess as to the value of x given the observed value y. If we knew the probability distribution of x given y, P(x|y), then we might want to pick the value of x that maximizes this distribution,

$$\hat{x} = \arg\max_{x} P(x \mid y).$$

Alternatively, if we want to minimize the mean squared error of our guesses, then we should pick the mean of $P(x \mid y)$:

$$\hat{x} = \int x P(x \mid y) dx.$$

So, if only we knew $P(x \mid y)$ then we could make an optimal guess.

Bayes' rule tells us how to calculate $P(x \mid y)$:

$$P(x \mid y) = \frac{P(y \mid x) P(x)}{P(y)}.$$

The two main things we need to specify here are $P(y|x)$ and $P(x)$. The first is easy, since we specified the noise, n, to be Gaussian, zero-mean, and additive. Thus,

$$P(y|x) = P(n+x|x)$$
$$= \frac{1}{\sqrt{2\pi}\sigma_n} e^{-\frac{(y-x)^2}{2\sigma_n^2}},$$

where σ_n^2 is the variance of the noise. For the prior, we have to draw upon our existing knowledge of x. Let's say x is the voltage of a car battery, and as an experienced mechanic you have observed the voltages on thousands of cars, using very accurate voltage meters, to have a mean of 12 volts, variance of 1 volt, with an approximate Gaussian distribution. Thus,

$$P(x) = \frac{1}{\sqrt{2\pi}} e^{-\frac{(x-12)^2}{2}}$$

Now we are in a position to write down the posterior on x:

$$P(x|y) \propto P(y|x)P(x)$$
$$= e^{-\frac{(y-x)^2}{2\sigma_n^2}} e^{-\frac{(x-12)^2}{2}}$$
$$= e^{-\frac{1}{2}\left[\frac{(y-x)^2}{2\sigma_n^2} + (x-12)^2\right]}.$$

The x which maximizes $P(x|y)$ is the same as that which minimizes the exponent in brackets which may be found by simple algebraic manipulation to be,

$$\hat{x} = \frac{y + 12\sigma_n^2}{1+\sigma_n^2}$$

The entire inference process is depicted graphically in terms of the probability distributions in figure:

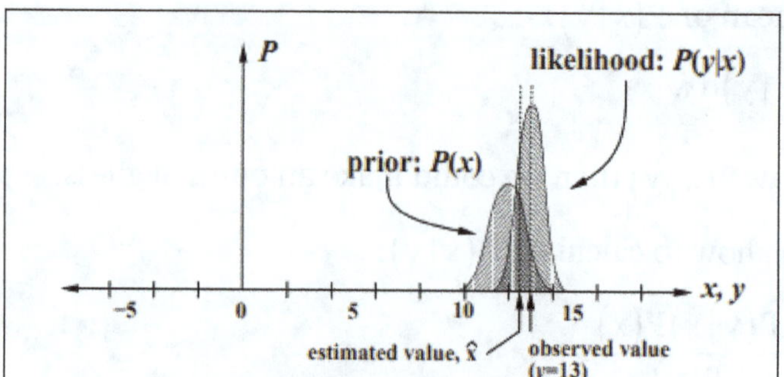

A Simple example of Bayesian interface.

Generative Models

Generative models, also known as "latent variable models" or "causal models," provide a way of modeling how a set of underlying causes. Such models have been commonly employed in the social sciences, usually in the guise of factor analysis, to make inferences about the causes leading to various social conditions or personality traits. More recently, generative models have been used to model the function of the cerebral cortex. The reason for this is that the cortex can be seen as solving a very complex inference problem, where it must select the best hypothesis for "what's out there" based on the massive data stream (gigabits per second) present on the sensory receptors.

The basic idea behind a generative model is illustrated in figure below. A set of multivariate data, D, is explained in terms of a set of underlying causes, α. For instance, the data on the left may be symptoms of a patient, the causes on the right might be various diseases, and the links would represent how the diseases give rise to the symptoms and how the diseases interact with each other. Alternatively, the data may be a retinal image array, in which case the causes would be the objects present in the scene along with their positions and that of the lighting source, and the links would represent the rendering operation for creating an image from these causes. In general the links may be linear (as is the case in factor analysis), or more generally they may instantiate highly non-linear interactions among the causes or between the causes and the data.

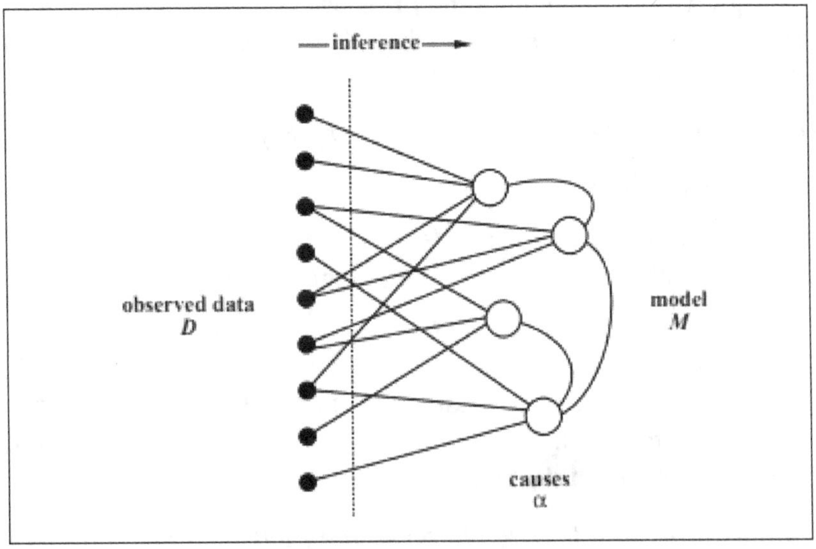

A generative model.

There are two fundamental problems to solve in a generative model. One is to infer the best set of causes to represent a specific data item, D_i . The other is to learn the best model, M, for explaining the entire set of data, $D = \{D_1, D_2, ..., D_n\}$. For example, each Di might correspond to a specific image, and D would represent the set of all observed images in nature. In modeling the operations of the cortex, the first problem may be seen as one of perception, while the second is one of adaptation.

Inference (Perception)

Inferring the best set of causes to explain a given piece of data usually involves maximizing the posterior over α (or alternatively computing its mean), similar to the computation of x in the simplified example.

$$\begin{aligned}\hat{\alpha} &= \arg\max_{\alpha} P(\alpha \mid D_i, M) \\ &= \arg\max_{\alpha} P(D_i \mid \alpha, M) P(\alpha \mid M).\end{aligned}$$

Note that the denominator $P(D_i \mid M)$ may be dropped here since D_i is fixed. All quantities are conditioned on the model, M, which specifies the overall "architecture" (such as depicted in the figure) within which the causes, α, are defined.

Learning (Adaptation)

The model, M, specifies the set of potential causes, their prior probabilities, and the generative process by which they give rise to the data. Learning a specific model, M, that best accounts for all the data is accomplished by maximizing the posterior distribution over the models, which according to Bayes' rule is,

$$P(M \mid D) \propto P(D \mid M) P(M).$$

Oftentimes though we are agnostic in the prior over the model, and so we may simply choose the model that maximizes the likelihood, $P(D \mid M)$. The total probability of all the data under the model is,

$$\begin{aligned}P(D \mid M) &= P(D_1 \mid M) \times P(D_2 \mid M) \times \ldots \times P(D_n \mid M) \\ &= \Pi_i P(D_i \mid M)\end{aligned}$$

where D_i denotes an individual data item (e.g., a particular image). The probability of an individual data item is obtained by summating over all of the possible causes for the data,

$$P(D_i \mid M) = \sum_{\alpha} P(D_i \mid \alpha, M) P(\alpha \mid M).$$

It is this summation that forms the most computationally formidable aspect of learning in the Bayesian framework. Usually there are ways we can approximate this sum though, and much effort goes into this step.

One typically maximizes the log-likelihood of the model for reasons stemming from information theory as well as the fact that many probability distributions are naturally expressed as exponentials. In this case, we seek a model, M^*, such that,

$$\begin{aligned}M^* &= \arg\max_{M} \log P(D \mid M) \\ &= \arg\max_{M} \sum_i \log P(D_i \mid M) \\ &= \arg\max_{M} \langle \log P(P \mid M) \rangle\end{aligned}$$

An example of where this approach has been applied in modeling the cortex is in understanding the receptive field properties of so-called simple cells, which are found in the primary visual cortex of all mammals. These cells have long been noted for their spatially localized, oriented, and bandpass receptive fields, but until recently there has been no explanation, in terms of a single quantitative theory, for why cells are built this way. In terms of Bayesian probability theory, one can understand the function of these cells as forming a model of natural images based on a linear superposition of sparse, statistically independent events. That is, if one utilizes a linear generative model where the prior on the causes is factorial and sparse, the set of linear weighting functions that emerge from the adaptation process above are localized, oriented, and bandpass, similar to the functions observed in cortical cells and also to the basic functions of commonly used wavelet transforms.

Frequentist Probability

If the outcome of an event is observed in a large number of independent repetitions of the event under roughly the same conditions, then it is a fact of the real world that the frequency of the outcome stabilizes as the number of repetitions increases. If another long sequence of such events is observed, the frequency of the outcome will typically be approximately the same as it was in the first sequence.

Unfortunately, the real world is not very tidy. For this reason it was necessary in the above statement to insert several weasel words. The use of 'roughly the same,' 'typically,' 'approximately,' and 'long sequence' make it clear that the stability phenomenon being described cannot be stated very precisely. A much clearer statement is possible within a mathematical model of this phenomenon. This discovery is due to Jacob Bernoulli who raised the following question.

The degree to which the frequency of an observed event varies about the probability of the event decreases as the number of events increases. He goes on to say that an important question that has never been asked before concerns the behavior of this variability as the number n of events increases indefinitely. He envisages two possibilities:

- As n gets larger and larger, the variability eventually shrinks to zero, so that for sufficiently large n the frequency will essentially pinpoint the probability p of the outcome.
- Alternatively, it is conceivable that there is a positive lower bound below which the vari ability can never fall so that p will always be surrounded by a cloud of uncertainty, no matter how large a number of events we observe.

Bernoulli then proceeds to prove the law of large numbers, which shows that it is (a) rather than (b) that pertains. More precisely, he proves that for any $a > 0$,

$$\Pr\left(\left|\frac{X}{n} - p\right| < a\right) \to 1 \text{ as } n \to \infty$$

where X/n is the frequency under consideration.

It is easy to be misled into the belief that this theorem proves something about the behavior of frequencies in the real world. It does not. The result is only concerned with properties of the mathematical model. What it does show is that the behavior of frequencies in the model is mirrored in a way that is much neater and more precise.

In fact, a result for the model much more precise than equation,

$$\Pr\left(\left|\frac{X}{n}-p\right|<a\right)\to 1 \text{ as } n\to\infty$$

was obtained by De Moivre. It gives the normal approximation,

$$\Pr\left(\left|\frac{X}{n}-p\right|<c/\sqrt{n}\right)\to\int_{-c/\sqrt{pq}}^{c/\sqrt{pq}}\varphi(x)dx$$

where φ denotes the standard normal density. This is again a theorem in the model. The approximate corresponding real-world phenomenon can be seen, for example, by observing a quincunx, a mechanical device using balls falling through 'random' paths to generate a histogram.

De Moivre's result was given a far reaching generalization by Laplace in the central limit theorem (CL T) concerning the behavior of the average $\bar{X}$ of n identically, independently distributed random variable $X_1,..., X_n$ with mean ξ and finite variance σ^2. It shows that,

$$\Pr\left(\left|\bar{X}-\xi\right|<c/\sqrt{n}\right)\to\int_{-c/\sigma}^{c/\sigma}\varphi(x)dx.$$

This reduces to equation above when X takes on the values of 1 and 0 with probabilities p and q, respectively. The CLT formed the basis of most frequentist inference throughout the nineteenth century.

The first systematic discussion of the frequentist approach was given by Venn, and an axiomatization based on frequencies in infinite random sequences (Kollectives) was attempted by von Mises. Because of technical difficulties his concept of a random sequence was modified by Solomonoff, Martin-Lof, and Kolmogorov, with the introduction of computational complexity. (An entirely different axiomatization based on events and their probabilities rather than random sequences was put forward by Kolmogorov, and has successfully served as a basis for both frequentist and subjective interpretations of probability).

The frequentist concept of probability described so far has met considerable criticism. One of the main objections is that it is not applicable to many situations to which one might want to apply probability assessments. Consider the following three possibilities:

- An actual sequence of repetitions may be available; for example, a sequence of coin tosses or a sequence of independent measurements of the same quantity.

- A sequence of repetitions may be available in principle but not likely to be carried out in practice; for example, the polio experiment of 1954 involving a sample of over a million children.
- A unique event which by its very nature can never be replicated, such as the outcome of a particular historical event; for example, whether a particular president will survive an impeachment trial. The conditions of this experiment cannot be duplicated.

The frequentist concept of probability can be applied in first and second case but not in the third situation. An alternative approach to probability which is applicable in all cases is the notion of probability as degree of belief; i.e., of a state of mind. The inference methods based on these two interpretations of the meaning of probability are called frequentist and Bayesian, respectively.

Although frequentist probability is considered objective, it has the following subjective feature. Its impact on a particular person will differ from one person to another. One patient facing a surgical procedure with a 1 percent mortality rate will consider this a dire prospect and emphasize the possibility of a fatal outcome. Another will shrug it off as so rare as not to be worth worrying about.

There exists a class of situations in which both approaches will lead to the same probability assessment. Suppose there is complete symmetry between the various outcomes; for example, in random sampling which is performed so that the drawing favors no sample over another. Then we expect the frequencies of the various outcomes to be roughly the same and will also, in our beliefs, assign the same probability to each of them.

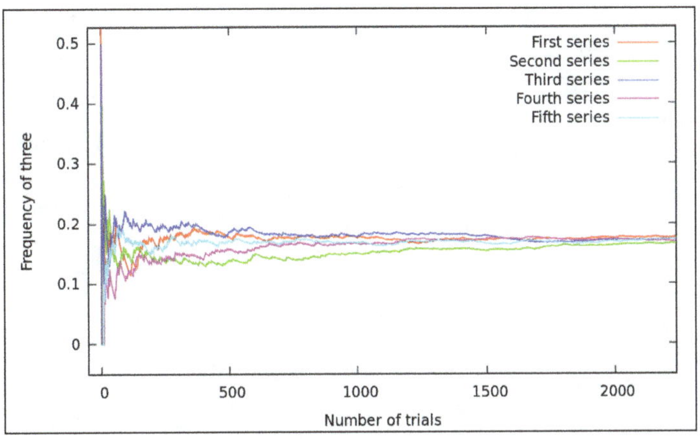

Let us now turn to a second criticism of frequentist probability. This concerns the difficulty of specifying what is meant by a repetition in the first sentence of this section. Consider once more the surgical procedure with 1 percent fatalities. This figure may represent the experience of thousands of cases, with the operation performed by different surgeons in different hospitals and- of course- on different patients. The rate of fatalities may vary from one hospital or surgeon to another and may, in particular, vary drastically with the condition, for example, the age and general health, of the patient.

Suppose a young woman requires this operation although her general health is very good. The frequency of a fatal outcome with patients sharing these characteristics may be much lower, and the 1 percent figure in that sense would be quite misleading for her. And yet she might be considered to have been obtained under 'roughly the same conditions,' namely to be drawn at random from the total population of persons requiring this surgery. To obtain the most useful figures one should identify the most important variables, classify the cases accordingly (for example, young, middle-aged, old; male, female; etc.) and then provide the frequency for each class. They will, of course, be meaningful only for the classes which contain a reasonable number of cases.

Probability Theory 2

- Probability Axioms
- Probability Space
- Sample Space
- Event in Probability Theory
- Mutual Exclusivity
- Boole's Inequality

The branch of mathematics that is associated with probability is termed as probability theory. Probability space, elementary event, complementary event, independent event, mutual exclusivity, Boole's inequality, etc. are a few of its aspects. This is an introductory chapter which will briefly introduce about probability theory.

A mathematical science in which the probabilities of certain random events are used to deduce the probabilities of other random events which are connected with the former events in some manner.

A statement to the effect that the probability of occurrence of a certain event is, say, 1/2, is not in itself valuable, since one is interested in reliable knowledge. Only results which state that the probability of occurrence of a certain event A is quite near to one or (which is the same thing) that the probability of the event not occurring is very small, represent ultimately valuable information. In accordance with the principle of "discarding sufficiently small probabilities", such an event is considered to be practically certain. It will be shown below that conclusions of scientific and practical interest are usually based on the assumption that the occurrence or non-occurrence of an event A depends on a large number of random factors, which are interconnected only to a minor extent. It may also be said, accordingly, that probability theory is the mathematical science of the laws governing the interaction of a large number of random factors.

The Subject Matter of Probability Theory

In order to describe a regular connection between certain conditions S and an event A the occurrence or non-occurrence of which can be established exactly, one of the following two schemes are usually employed in science:

- The occurrence of event A follows each realization of the conditions S This is the form of, say, all the laws of classical mechanics which state that under given initial conditions and forces acting on a body or a system of bodies, the motion will proceed in a uniquely determined manner.
- Under the conditions S the occurrence of event A has a definite probability $P(A|S)$ which is equal to p. For instance, the laws governing ionizing radiation say that, for each radioactive substance there is a definite probability that, in a given period of time, some number N of the atoms of the substance will decay.

The frequency of occurrence of event A in a given sequence of n trials (i.e. n repeated realizations of the conditions S) is the ratio $p = m/n$ between the number m of trials in which A has occurred to the total number of trials n. That there is in fact a definite probability p for A to occur, under the conditions S, is manifested by the fact that in almost-all sufficiently large sequences of trials the frequency of occurrence of A is approximately equal to p. Any mathematical model which is intended to be a schematic description of the connection between conditions S and a random event A, usually also contains certain assumptions about the nature and the degree of dependence of the trials. After these additional assumptions have been made, it is possible to give a quantitative, more precise expression of the somewhat vague statement made above to the effect that the frequency is close to the probability.

Statistical relationships, were first noted for games of chance such as throwing a die. Statistical relationships concerning births and deaths have been known for a very long time (e.g. the probability of a newborn (human) baby being a boy is 0.515). The end of the 19th century and the first half of the 20th century have witnessed the discovery of a large number of statistical laws in physics, chemistry, biology, and other sciences. It should be noted that statistical laws are also involved in schemes not directly related to the concept of randomness, e.g. in the distribution of digits in tables of functions, etc. This fact is utilized, in particular, in the "simulation" of random phenomena.

That methods of probability theory can be used in studying the relationships prevailing in a large number of sciences apparently unrelated to each other is due to the fact that probabilities of occurrence of events invariably satisfy certain simple laws. The study of the properties of the probability of occurrence of events, based on these simple laws, forms the subject matter of probability theory.

Fundamental Concepts in Probability Theory

The fundamental concepts in probability theory, as a mathematical discipline, are most simply exemplified within the framework of so-called elementary probability theory. Each trial T considered in elementary probability theory is such that it yields one and only one outcome or, as it is called, one of the elementary events $\omega 1,...,\omega_S$, which are supposed to be finite in number. To each outcome ω_k a non-negative number p_k is connected — the probability of this outcome. The sum of the numbers p_k must be one. Consider events A which are characterized by the condition.

“either wi or wj... or wk occurs.”

The outcomes $\omega_i, \omega_j, ..., \omega_k$ are said to be favourable to A and, by definition, one says that the probability $P(A)$ of A is equal to the sum of the probabilities of the outcomes favourable to this event:

$$P(A) = p_i + p_j + ... + p_k$$

If there are r outcomes favourable to A, then the special case $p_i = ... = p_s = 1/s$ yields the formula:

$$P(A) = \frac{r}{s}.$$

Formula $P(A) = \frac{r}{s}$ expresses the so-called classical concept of probability, according to which the probability of some event A is equal to the ratio between the number r of outcomes favourable to A and the number s of all “equally probable” outcomes. The computation of probabilities is thus reduced to counting the number of outcomes favourable to A and often proves to be a difficult problem in combinatorics.

Example: Each one of the 36 possible outcomes of throwing a pair of dice may be denoted by (i, j), where i is the number of dots shown by the first die, while j is the number of dots shown by the second. Event A — “the sum of the dots is 4” — is favoured by three outcomes: $(1;3),(2;2),(3;1)$ Thus, $P(A) = 3/36 = 1/12$.

The problem of determining the numerical values of the probabilities p_k in a given specific problem lies, strictly speaking, outside the scope of probability theory as a discipline of pure mathematics. In some cases these values are established as a result of processing the results of a large number of observations. In other cases it is possible to predict the probabilities of encountering given events in a given trial theoretically. Such a prediction is frequently based on an objective symmetry of the connections between the conditions under which the trial is conducted and the outcomes of the trials, and in such cases leads to a formula like $P(A) = \frac{r}{s}$. Let, for instance, the trial consist in throwing a die in the form of a cube made of a homogeneous material.

One may then assume that each side of the die has a probability of 1/6 of "coming out". In this case the assumption that all outcomes are equally probable is confirmed by experiment. Examples of this kind in fact form the basis of the classical definition of a probability.

A more detailed and thorough explanation for the causes of equal probabilities of individual outcomes in some special cases may be given by the so-called method of arbitrary functions. The method is explained below by taking again dice throwing as an example. Let the conditions of the trials be such that accidental effects of air on the die are negligible. In such a case, if the initial position, the initial velocity and the mechanical properties of the die are known exactly, the motion of the die may be calculated by the methods of classical mechanics, and the result of the trial may be reliably predicted. In practice, the initial conditions can never be determined with absolute accuracy and even very small changes in the initial velocity will produce a different result, provided the period of time t between the throw and the fall of the die is sufficiently long. It has been found that, under very general assumptions with respect to the probability distribution of the initial values (hence the name of the method), the probability of each one of the six possible outcomes tends to 1/6 as $t \to \infty$.

A second example consists of the shuffling of a pack of cards in order to ensure that all possible distributions are equally probable. Here, the transition from one distribution of the cards to the next as a result of two successive shuffles is usually random. The tendency to equi-probability is established by methods of the theory of Markov chains.

Both these cases can be seen as part of general ergodic theory.

Given a certain number of events, two new events may be defined: their union (sum) and combination (product, intersection). The event B: "at least one of $A_1 \ldots A_r$ occurs", is said to be the union of events $A_1, \ldots, A_r$.

The event C: "$A_1 \ldots$ and A_r occur", is said to be the combination or intersection of events $A_1, \ldots, A_r$.

The symbols for union and intersection of events are $\cup$ and $\cap$, respectively. Thus:

$$B = A_1 \cup \ldots \cup A_r, \quad C = A_1 \cap \ldots \cap A_r$$

Two events A and B are said to be mutually exclusive if their joint occurrence is impossible, i.e. if none of the possible results of a trial favours both A and B. If the events A_i are identified with the sets of their favourable outcomes, events B and C will be identical with the union and the intersection of the respective sets.

Two fundamental theorems in probability theory — theorems on addition and multiplication of probabilities — are connected with the operations just introduced.

The theorem on addition of probabilities. If the events $A_1,...,A_r$ are such that any two of them are mutually exclusive, the probability of their union is equal to the sum of their probabilities.

Thus, "the sum of the dots is 4 or less" is the sum of the three mutually exclusive events A_2, A_3, A_4 in which the sum of the dots is 2, 3 and 4, respectively. The probabilities of these events are 1/36, 2/36 and 3/36, respectively; in accordance with the addition theorem, $P(B)$ is equal to,

$$\frac{1}{36}+\frac{2}{36}+\frac{3}{36}=\frac{6}{36}=\frac{1}{6}$$

The conditional probability of event B occurring if condition A is met is defined by the formula:

$$P(B|A)=\frac{P(A\cap B)}{P(A)},$$

which may be shown to be in complete agreement with the properties of the frequencies of occurrence. Events $A_1,...,A_r$. are said to be independent if the conditional probability of any one of the events occurring under the condition that some of the other events have also occurred is equal to its "unconditional" probability.

The theorem on multiplication of probabilities. The probability of joint occurrence of events $A_1,...,A_r$. is equal to the probability of occurrence of event A_1 multiplied by the probability of occurrence of event A_2 on the condition that A_1 has in fact occurred,...., multiplied by the probability of occurrence of event A_r on the condition that the events $A_1,...,A_r$. have in fact occurred. If the events are independent, the multiplication theorem yields the formula:

$$P(A_1\cap...\cap A_r)=P(A_1)...P(A_r),$$

i.e. the probability of joint occurrence of independent events is equal to the product of the probabilities of these events. Formula $P(A_1\cap...\cap A_r)=P(A_1)...P(A_r)$ remains valid if some of the events are replaced in both its parts by the complementary events.

Example: Four shots are fired at a target, the probability of hitting the target being 0.2 with each shot. The hits scored in different shots are considered to be independent events. What will be the probability of hitting the target exactly three times?

Each outcome of a trial can be symbolized by a sequence of four letters (e.g. (h,m,m,h) means that the first and fourth shots were hits, while the second and the third shots were misses). The total number of outcomes will be $2\times2\times2=16$. Since the results of individual shots are assumed to be independent, the probability of the outcomes must be determined with the aid of formula $P(A_1 \cap ... \cap A_r) = P(A_1)...P(A_r)$ including the comment which accompanies it. Thus, the probability of the outcome (h, m, m, m) will be,

$$0.20.80.80.8 = 0.1024$$

where $0.8 = 1 - 0.2$ is the probability of miss in a single shot. The outcomes favouring the event "the target is hit three times" are (h, h, h, m), (h, h, m, h), (h, m, h, h), and (m, h, h, h). The probabilities of all four outcomes are equal:

$$0.2.0.2.0.2.0.8 = ... = 0.8.0.2.0.2.0.2 = 0.0064$$

so that the probability of the event is,

$$4.0.0064 = 0.256.$$

A generalization of the above reasoning leads to one of the fundamental formulas in probability theory: If the events $A_1, ..., A_n$ are independent and if the probability of each individual event occurring is p, then the probability of occurrence of exactly m such events is,

$$P_n(m) = C_n^m p^m (1 - p)^{n-m},$$

where $C_n^m = (\square_m^n)$ denotes the number of combinations of m elements out of n elements. If n is large, computations according to formula $P_n(m) = C_n^m p^m (1 - p)^{n-m}$ become laborious. In the above example, let the number of shots be 100; one has to find the probability x of the number of hits being between 8 and 32. The use of formula $P_n(m) = C_n^m p^m (1 - p)^{n-m}$ and of the addition theorem yields an accurate, but unwieldy expression for the probability value sought, namely:

$$x = \sum_{m=8}^{32} \binom{100}{m} (0.2)^m (0.8)^{100-m}$$

An approximate value of the probability x may be found by the use of the Laplace theorem:

$$x \approx \frac{\sim 1}{\sqrt{2\pi}} \int_{-3}^{+3} e^{-z^2/2} dz = 0.9973$$

the error not exceeding 0.0009. This result shows that the occurrence of the event $8 \le m \le 32$ is practically certain. This is a very simple, but typical, example of the use of limit theorems in probability theory.

Another fundamental formula in elementary probability theory is the so-called formula of total probability: If events $A_1,\ldots,A_r$ are pairwise mutually exclusive and if their union is the sure event, the probability of any single event B is equal to the sum:

$$P(B)=\sum_{k=1}^{r}P(B\,|\,A_k)P(A_k).$$

The theorem on multiplication of probabilities is particularly useful when compound trials are considered. One says that a trial T is composed of trials $T_1,\ldots,$ if each outcome of T is a combination of certain outcomes $A_i,B_j,\ldots,X_k,Y_l$ of the respective trials $T_1,T_2,\ldots,T_{n-1},T_n$. Frequently one is in the situation where the probabilities,

$$P(A_i),P(B_j\,|\,A_i),\ldots,P(Y_l\,|\,A_i\cap\ldots\cap X_k)$$

are, for some reason, known. The data in $P(A_i),P(B_j\,|\,A_i),\ldots,P(Y_l\,|\,A_i\cap\ldots\cap X_k)$ together with the multiplication theorem may then be used to determine the probabilities $P(E)$ for all outcomes E of the compound trial, as well as the probabilities of all events connected with this trial. Two types of compound trials are especially important in practice: A) the individual trials are independent, i.e. the probabilities in $P(A_i),P(B_j\,|\,A_i),\ldots,P(Y_l\,|\,A_i\cap\ldots\cap X_k)$ are equal to the unconditional probabilities $P(A_i),P(B_j),\ldots,P(X_k),P(y_l)$; B) the probabilities of the outcomes of a given trial are only affected by the outcomes of the immediately preceding trial, i.e. the probabilities in $P(A_i),P(B_j\,|\,A_i),\ldots,P(Y_l\,|\,A_i\cap\ldots\cap X_k)$ are equal, respectively, to $P(A_i),P(B_j\,|\,A_i),\ldots,P(Y_l\,|\,X_k)$. One then says that the trials are connected in a Markov chain. The probabilities of all events connected with a compound trial are here fully determined by the initial probabilities $P(A_i)$ and by the intermediate probabilities $P(A_i),P(B_j\,|\,A_i),\ldots,P(Y_l\,|\,X_k)$.

Random variables: If each outcome of a trial T is put into correspondence with a number x_r, one says that a random variable X has been specified. Among the numbers $x_1,\ldots,x_s$ there may be equals; the set of different values of x_r, where $r=1,\ldots,s$, is the set of possible values of the random variable. The set of possible values of a random variable, together with their respective probabilities is said to be the probability distribution of the random variable. Thus, in the example of throwing a pair of dice, to each outcome (i,j) of the trial there corresponds the value of the random variable $X=i+j$ which is the sum of the dots on the two dice. The possible values are $2,3,\ldots,12$ and their respective probabilities are $1/36,2/36,\ldots,1/36$.

In a joint study of several random variables one introduces the concept of their joint distribution, which is defined by indicating the possible values of each one, and the probabilities of joint occurrence of the events,

$$\{X_1=x_1\},\ldots,\{X_n=x_n\},$$

where x_k is one of the possible values of the variable X_k. Random variables are said to be independent if the events in $\{X_1 = x_1\},\ldots,\{X_n = x_n\}$ are independent whatever the choice of the x_k. The joint distribution of random variables can be used to calculate the probability of any event defined by these variables, e.g. of the event etc.

$$a < X_1 + \ldots + X_n < b,$$

Often, instead of giving the distribution of a random variable completely, one uses a, not too large, collection of numerical characteristics. The ones most often used are the mathematical expectation and the dispersion (variance).

The fundamental characteristics of a joint distribution of several random variables include — in addition to the mathematical expectations and the variances of these variables — also the correlation coefficients, etc. The meaning of these characteristics can be made clear, to a considerable extent, by limit theorems.

The scheme of trials with a finite number of outcomes proves inadequate even in the simplest applications of probability theory. Thus, in the study of the random dispersion of the hitting sites of projectiles around the centre of a target, or in the study of random errors in the determination of some value, etc., it is not possible to limit the model to trials with a finite number of outcomes. Moreover, such outcomes may, in some cases, be expressed by a number or a set of numbers, while in other cases the outcome of a trial may be a function (e.g. a record of the variation of atmospheric pressure at a given location over a certain period of time), a set of functions, etc. It should be noted that many definitions and theorems given above, after suitable modifications, are also applicable in these more general cases, although the forms in which the probability distribution is presented are different. Here, the classical "equal probability of each outcome" is replaced by a uniform distribution of the objects under consideration in some area (this is exactly what is meant when speaking of a point randomly selected in a given area, a randomly selected tangent to some figure, etc.)

Major changes are introduced in the definition of a probability which, in the elementary case, is given by formula $P(A) = \frac{r}{s}$. In the more general schemes now discussed, the events are the union of an infinite number of elementary events the probability of each one of which may be zero. Thus, the property which is described by the addition theorem is not a consequence of the definition of probability, but is part of it.

The logical scheme of constructing the fundamentals of probability theory which is most often employed was developed in 1933 by A.N. Kolmogorov. The fundamental characteristics of this scheme are the following. In studying a real problem by the methods of probability theory, the first step is to isolate a set U of elements u, called elementary events. Any event can be fully described by the set of elementary events favourable to it, and is therefore considered as some set of elementary events. To some events A are assigned certain numbers $P(A)$, which are called their probabilities and which satisfy the following conditions:

- $0 \le P(A) \le 1$;
- $P(U) = 1$;
- if the events $A_1, \dots, A_n$ are pairwise mutually exclusive, and if A is their union, then,

  $P(A) = P(A_1) + \dots + P(A_n)$

(additivity of probabilities).

In order to construct a mathematically rigorous theory, the domain of definition of $P(A)$ must be a σ-algebra, and condition $P(A_i), P(B_j \mid A_i), \dots, P(Y_1 \mid A_i \cap \dots \cap X_k)$ must also be met for an infinite sequence of events which are mutually exclusive (countable additivity of probabilities). Non-negativity and countable additivity are fundamental properties of measures. Thus, probability theory may be formally regarded as a part of measure theory. The fundamental concepts of probability theory are then viewed in a new light: random variables become measurable functions, their mathematical expectations become the abstract integrals of Lebesgue, etc. However, the main problems of probability theory and of measure theory are different. In probability theory, the basic, specific concept is that of independence of events, trials and random variables. Moreover, probability theory comprises a thorough study of subjects such as probability distributions, conditional mathematical expectations, etc.

The following comments may be made on the scheme described above. In accordance with the scheme, each probability model is based on a probability space, which is a triplet (Ω, S, P), where Ω is a set of elementary events, S is a σ-algebra of subsets of Ω and P is a probability distribution (a countably-additive normalized measure) on S. Two achievements of this scheme are the definition of probabilities in infinite-dimensional spaces (in particular, in spaces connected with infinite sequences of trials and stochastic processes), and the general definition of conditional probabilities and conditional mathematical expectations (with respect to a given random variable, etc).

Subsequent development of probability theory showed that the above definition of a probability space can be expediently narrowed. These developments have led to concepts such as perfect distributions and probability spaces, Blackwell spaces, Radon probability measures on topological (linear) spaces, etc.

There are also other approaches to the fundamental concepts of probability theory, such as axiomatization, the principal object of which is a normalized Boolean algebra of events. Here, the principal advantage (provided that the algebra being considered is complete in the metric sense) consists of the fact that for any directed system of events the following relations are true:

$$P\left(\bigcup_{\alpha} A\alpha\right) = \sup_{\alpha} P(A\alpha), \qquad A\alpha \uparrow$$

$$P\left(\bigcap_{\alpha} A\alpha\right) = \inf_{\alpha} P(A\alpha), \qquad A\alpha \downarrow$$

It is possible to axiomatize the concept of a random variable as an element of some commutative algebra with a positive linear functional defined on it (the analogue of the mathematical expectation). This is the starting point for non-commutative and quantum probability.

Limit Theorems

In a formal exposition of probability theory limit theorems appear as a kind of superstructure over its elementary sections in which all problems are of a finite, purely arithmetical nature. However, the cognitive value of probability theory can only be revealed by these limit theorems. Thus, it is shown by the Bernoulli theorem that the frequency of occurrence of a given event in independent trials is usually close to its probability, while the Laplace theorem yields the probabilities of deviations of this frequency from its limiting value. In a similar manner, the meaning of the characteristics of a random variable such as its mathematical expectation and variance are explained by the law of large numbers and the central limit theorem.

Let,

$$X_1, \ldots X_n, \ldots,$$

be independent random variables with the same probability distribution, with $EX_k = a$, $DX_k = \sigma^2$, and let Y_n be the arithmetical average of the first n variables of the sequence $X_1, \ldots X_n, \ldots,$:

$$Y_n = \frac{X_1 + \ldots + X_n}{n}$$

In accordance with the law of large numbers, for any $\in > 0$ the probability of the inequality $|Y_n - a| \leq \in$ tends to one as $n \to \infty$, so that, as a rule, the value of Y_n is close to a. This result is rendered more precise by the central limit theorem, according to which the deviations of $\in > 0$ from a are approximately normally distributed, with mathematical expectation 0 and variance σ^2 / n. Thus, in order to calculate (to a first approximation) the probability of some deviation of Y_n from α for large n, there is no need to know the distribution of the variables X_n in all details; knowledge of their variance is sufficient. If a higher accuracy of approximation is required, moments of higher order must also be used.

The above statements, with suitable modifications, may be extended to random vectors (in finite-dimensional and in some infinite-dimensional spaces). The independence conditions may be replaced by conditions of a "weak" (in some sense) dependence of the X_n. Limit theorems of distributions on groups, of distributions of values of arithmetic functions, etc., are also known.

In applications — in particular, in mathematical statistics and statistical physics — it may be necessary to approximate small probabilities (i.e. probabilities of events of the type $|Y_n - a| > \epsilon$) with a high relative accuracy. This involves major corrections to the normal approximation.

It was noted in the nineteen twenties that quite natural non-normal limit distributions may appear even in schemes of sequences of uniformly-distributed and independent random variables. For instance, let X_1 be the time which elapses until some randomly varying variable has returned to its initial location, let X_2 be the time between the first and the second such returns, etc. Then, under very general conditions, the distribution of the sum $X_1 + ... + X_n$ (i.e. the time elapsing prior to the n-th return) will, after multiplication by $n^{-1/\alpha}$ (where α is a constant smaller than one), converge to some limit distribution. Thus, the time prior to the n-th return increases, roughly speaking, in proportion to $n^{1/\alpha}$, i.e. at a faster rate than n (if the law of large numbers were applicable, it would be of order n). This is seen in the case of a Bernoulli random walk (in which another paradoxical law — the arcsine law — also appears).

The principal method of proof of limit theorems is the method of characteristic functions and the related methods of Laplace transforms and of generating functions). In a number of cases it becomes necessary to invoke the theory of functions of a complex variable.

The mechanism of the existence of most limit relationships can be completely understood only in the context of the theory of stochastic processes.

Stochastic Processes

During the past few decades the need to consider stochastic processes — i.e. processes with a given probability of their proceeding in a certain manner, arose in certain physical and chemical investigations, along with the study of one-dimensional and higher-dimensional random variables. The coordinate of a particle executing a Brownian motion may serve as an example of a stochastic process. In probability theory a stochastic process is usually regarded as a one-parameter family of random variables $X(t)$. In most applications the parameter t is time, but it may also be an arbitrary variable, and in such cases it is usual to speak of a random function (if t is a point in space — a random field). If the parameter t runs through integer values, the random function is said to be a random sequence (or a time series). While a random variable may be characterized by a distribution law, a stochastic process may be characterized by the totality of joint distribution laws for $X(t_1),...,X(t_n)$ for all possible moments of time $t_1,...,t_n$ for any $n>0$ (the so-called finite-dimensional distributions). The most interesting concrete results in the theory of stochastic processes were obtained in two fields — Markov processes and stationary stochastic processes; the interest in martingales is now also strongly increasing.

Chronologically, Markov processes were the first to be studied. A stochastic process $X(t)$ is said to be a Markov process if, for any two moments of time t_0 and t_1 $((t_0<t_1))$, the conditional probability distribution of $X(t_1)$ depends, provided all values of $X(t)$ for $t \le t_0$ are given, only on $X(t_0)$. For this reason Markov processes are sometimes referred to as processes without after-effect. Markov processes are a natural extension of the deterministic processes studied in classical physics. In deterministic processes the state of the system at the moment of time t_0 uniquely determines the course of the process in the future; in Markov processes the state of the system at the moment of time t_0 uniquely determines the probability distribution of the course of the process at $t>t_0$, and this distribution cannot be altered by any information on the course of the process prior to the moment of time t_0.

Just as the study of continuous deterministic processes is reduced to differential equations involving functions which describe the state of the system, the study of continuous Markov processes can, to a large extent, be reduced to differential or differential-integral equations with respect to the distribution of the probabilities of the process.

Another major subject in the field of stochastic processes is the theory of stationary stochastic processes. The stationary nature of a process, i.e. the fact that its probability relations remain unchanged with time, imposes major restrictions on the process and makes it possible to arrive at several important deductions based on this premise.

A major part of the theory is based only on the assumption of stationarity in a wide sense, viz. that the mathematical expectations $EX(t)$ and $EX(t)X(t+\tau)$ are independent of t. This assumption leads to the so-called spectral decomposition:

$$X(t)=\int_{-\infty}^{+\infty} e^{it\lambda}dz(\lambda)$$

where $z(\lambda)$ is a random function with uncorrelated increments. Methods of best (in the mean square) linear interpolation, extrapolation and filtering have been developed for stationary processes.

Recently a rather large class of processes, the so-called semi-martingales, which serves to solve problems of optimal non-linear filtering, interpolation and extrapolation, has been isolated. A substantial part of the relevant analytical apparatus is provided by stochastic differential equations, stochastic integrals and martingales. A distinguishing feature of a martingale $X(t)$ is the fact that the conditional mathematical expectation of $X(t)$ is $X(s)$, given the values of $X(u)$ for $u \le s, s<t$.

The theory of stochastic processes is closely connected with the classical problems on limit theorems for sums of random variables. Distributions which appear as limit distributions in the study of sums of random variables become exact distributions of appropriate characteristics in the theory of stochastic processes. This fact makes it possible to demonstrate many limit theorems with the aid of these associated stochastic processes.

One may finally note that the logically unobjectionable definition of the concepts connected with stochastic processes within the framework of the axiomatics discussed above has always presented and still presents a large number of difficulties of measure-theoretic nature. These are connected, for example, with the definition of probabilistic continuity, differentiability, etc. of stochastic processes. This is why monographs on the theory of stochastic processes devote about half their space to the analysis of the development of measure-theoretic constructions.

Probability Axioms

One strategy in mathematics is to start with a few statements, then build up more mathematics from these statements. The beginning statements are known as axioms. An axiom is typically something that is mathematically self-evident. From a relatively short list of axioms, deductive logic is used to prove other statements, called theorems or propositions.

The area of mathematics known as probability is no different. Probability can be reduced to three axioms. This was first done by the mathematician Andrei Kolmogorov. The handful of axioms that are underlying probability can be used to deduce all sorts of results. But what are these probability axioms?

We suppose that we have a set of outcomes called the sample space *S*. This sample space can be thought of as the universal set for the situation that we are studying. The sample space is comprised of subsets called events $E_1, E_2, \ldots, E_n$.

We also assume that there is a way of assigning a probability to any event *E*. This can be thought of as a function that has a set for an input, and a real number as an output. The probability of the event *E* is denoted by *P*(*E*).

Axiom One

The first axiom of probability is that the probability of any event is a nonnegative real number. This means that the smallest that a probability can ever be is zero and that it cannot be infinite. The set of numbers that we may use are real numbers. This refers to both rational numbers, also known as fractions, and irrational numbers that cannot be written as fractions.

One thing to note is that this axiom says nothing about how large the probability of an event can be. The axiom does eliminate the possibility of negative probabilities. It reflects the notion that smallest probability, reserved for impossible events, is zero.

Axiom Two

The second axiom of probability is that the probability of the entire sample space is one. Symbolically we write $P(S) = 1$. Implicit in this axiom is the notion that the sample space is everything possible for our probability experiment and that there are no events outside of the sample space.

By itself, this axiom does not set an upper limit on the probabilities of events that are not the entire sample space. It does reflect that something with absolute certainty has a probability of 100%.

Axiom Three

The third axiom of probability deals with mutually exclusive events. If E_1 and E_2 are mutually exclusive, meaning that they have an empty intersection and we use U to denote the union, then $P(E_1 \mathrm{U} E_2) = P(E_1) + P(E_2)$.

The axiom actually covers the situation with several (even countably infinite) events, every pair of which are mutually exclusive. As long as this occurs, the probability of the union of the events is the same as the sum of the probabilities:

$$P(E_1 \mathrm{U} E_2 \mathrm{U} ... \mathrm{U} E_n) = P(E_1) + P(E_2) + ... + E_n$$

Although this third axiom might not appear that useful, we will see that combined with the other two axioms it is quite powerful indeed.

Axiom Applications

The three axioms set an upper bound for the probability of any event. We denote the complement of the event E by E^C. From set theory, E and E^C have an empty intersection and are mutually exclusive. Furthermore $E \mathrm{U} E^C = S$, the entire sample space.

These facts, combined with the axioms give us:

$$1 = P(S) = P(E \mathrm{U} E^C) = P(E) + P(E^C)$$

We rearrange the above equation and see that $P(E) = 1 - P(E^C)$. Since we know that probabilities must be nonnegative, we now have that an upper bound for the probability of any event is 1.

By rearranging the formula again we have $P(E^C) = 1 - P(E)$. We also can deduce from this formula that the probability of an event not occurring is one minus the probability that it does occur.

The above equation also provides us a way to calculate the probability of the impossible event, denoted by the empty set. To see this, recall that the empty set is the complement of the universal set, in this case S^C. Since $1 = P(S) + P(S^C) = 1 + P(S^C)$, by algebra we have $P(S^C) = 0$.

Consequences

From the Kolmogorov axioms, one can deduce other useful rules for studying probabilities. The proofs of these rules are a very insightful procedure that illustrates the power the third axiom, and its interaction with the remaining two axioms. Four of the immediate corollaries and their proofs are shown below:

Monotonicity

$$\text{if} \quad A \subseteq B \quad \text{then} \quad P(A) \leq P(B).$$

If A is a subset of, or equal to B, then the probability of A is less than, or equal to the probability of B.

Proof of Monotonicity

In order to verify the monotonicity property, we set $E_1 = A$ and $E_2 = B \setminus A$, where $A \subseteq B$ and $E_i = \varnothing$ for $i \geq 3$. It is easy to see that the sets E_i are pairwise disjoint and $E_1 \cup E_2 \cup \cdots = B$. Hence, we obtain from the third axiom that,

$$(A) + P(B \setminus A) + \sum_{i=3}^{\infty} P(E_i) = P(B).$$

Since, by the first axiom, the left-hand side of this equation is a series of non-negative numbers, and since it converges to $P(B)$ which is finite, we obtain both $P(A) \leq P(B)$ and $P(\varnothing) = 0$.

The Probability of the Empty Set

$$P(\varnothing) = 0.$$

In some cases, $\varnothing$ is not the only event with probability 0.

Proof of Probability of the Empty Set

As shown in the previous proof, $P(\varnothing) = 0.$. However, this statement is seen by contradiction: if $P(\varnothing) = a$ then the left hand side $[P(A) + P(B \setminus A) + \sum_{i=3}^{\infty} P(E_i)]$ is not less than infinity; $\sum_{i=3}^{\infty} P(E_i) = \sum_{i=3}^{\infty} P(\varnothing) = \sum_{i=3}^{\infty} a = \begin{cases} 0 & \text{if } a = 0, \\ \infty & \text{if } a > 0. \end{cases}$

If $a > 0$ then we obtain a contradiction, because the sum does not exceed $P(B)$ which is finite. Thus, $a = 0$. We have shown as a byproduct of the proof of monotonicity that $P(\varnothing) = 0$.

The Complement Rule

$$P(A^c) = P(\Omega \setminus A) = 1 - P(A)$$

Proof of the Complement Rule

Given A and A^c are mutually exclusive and that $A \cup A^c = \Omega$:

$P(A \cup A^c) = P(A) + P(A^c)$... (by axiom 3)

and, $P(A \cup A^c) = P(\Omega) = 1$... (by axiom 2)

$$\Rightarrow P(A) + P(A^c) = 1$$

$$\therefore P(A^c) = 1 - P(A)$$

The Numeric Bound

It immediately follows from the monotonicity property that,

$$0 \le P(E) \le 1 \qquad \forall E \in F.$$

Proof of the Numeric Bound

Given the complement rule $P(E^c) = 1 - P(E)$ and axiom 1 $P(E^c) \ge 0$:

$$1 - P(E) \ge 0$$
$$\Rightarrow 1 \ge P(E)$$
$$\therefore 0 \le P(E) \le 1$$

Further Consequences

Another important property is:

$$P(A \cup B) = P(A) + P(B) - P(A \cap B).$$

This is called the addition law of probability, or the sum rule. That is, the probability that

A or B will happen is the sum of the probabilities that *A* will happen and that *B* will happen, minus the probability that both *A and B* will happen. The proof of this is as follows:

Firstly,

$$P(A \cup B) = P(A) + P(B \setminus A) \ldots \textit{(by Axiom 3)}$$

So,

$$P(A \cup B) = P(A) + P(B \setminus (A \cap B)) \text{ (by } B \setminus A = B \setminus (A \cap B)\text{)}.$$

Also,

$$P(B) = P(B \setminus (A \cap B)) + P(A \cap B)$$

and eliminating $P(B \setminus (A \cap B))$ from both equations gives us the desired result.

An extension of the addition law to any number of sets is the inclusion–exclusion principle.

Setting *B* to the complement A^c of *A* in the addition law gives

$$P\left(A^c\right) = P(\Omega \setminus A) = 1 - P(A)$$

That is, the probability that any event will *not* happen (or the event's complement) is 1 minus the probability that it will.

Simple Example: Coin Toss

Consider a single coin-toss, and assume that the coin will either land heads (H) or tails (T) (but not both). No assumption is made as to whether the coin is fair.

We may define:

$$\Omega = \{H, T\}$$

$$F = \{\varnothing, \{H\}, \{T\}, \{H, T\}\}$$

Kolmogorov's axioms imply that:

$$P(\varnothing) = 0$$

The probability of *neither* heads *nor* tails, is 0.

$$P(\{H, T\}^c) = 0$$

The probability of *either* heads *or* tails, is 1.

$$P(\{H\}) + P(\{T\}) = 1$$

The sum of the probability of heads and the probability of tails, is 1.

Probability Space

A probability space or a probability triple $(\Omega, \mathcal{F}, P)$is a mathematical construct that models a real-world process (or "experiment") consisting of states that occur randomly. A probability space is constructed with a specific kind of situation or experiment in mind. One proposes that each time a situation of that kind arises, the set of possible outcomes is the same and the probabilities are also the same.

A probability space consists of three parts:

- A sample space, Ω, which is the set of all possible outcomes.
- A set of events $\mathcal{F}$, where each event is a set containing zero or more outcomes.
- The assignment of probabilities to the events; that is, a function P from events to probabilities.

An outcome is the result of a single execution of the model. Since individual outcomes might be of little practical use, more complex *events* are used to characterize groups of outcomes. The collection of all such events is a σ-algebra $\mathcal{F}$. Finally, there is a need to specify each event's likelihood of happening. This is done using the *probability measure* function, P.

Once the probability space is established, it is assumed that "nature" makes its move and selects a single outcome, ω, from the sample space Ω. All the events in $\mathcal{F}$ that contain the selected outcome ω (recall that each event is a subset of Ω,) are said to "have occurred". The selection performed by nature is done in such a way that if the experiment were to be repeated an infinite number of times, the relative frequencies of occurrence of each of the events would coincide with thc probabilities prescribed by the function P.

The Russian mathematician Andrey Kolmogorov introduced the notion of probability space, together with other axioms of probability, in the 1930s. Nowadays alternative approaches for axiomatization of probability theory exist, e.g. algebra of random variables.

A probability space is a mathematical triplet $(\Omega, \mathcal{F}, P)$ that presents a model for a particular class of real-world situations. As with other models, its author ultimately defines which elements Ω, $\mathcal{F}$, and P will contain:

- The sample space Ω, is the set of all possible outcomes. An outcome is the result of a single execution of the model. Outcomes may be states of nature, possibilities, experimental results and the like. Every instance of the real-world situation (or run of the experiment) must produce exactly one outcome. If outcomes of different runs of an experiment differ in any way that matters, they are distinct outcomes. Which differences matter depends on the kind of analysis we want to do. This leads to different choices of sample space.

- The σ-algebra $\mathcal{F}$ is a collection of all the events we would like to consider. This collection may or may not include each of the elementary events. Here, an "event" is a set of zero or more outcomes, i.e., a subset of the sample space. An event is considered to have "happened" during an experiment when the outcome of the latter is an element of the event. Since the same outcome may be a member of many events, it is possible for many events to have happened given a single outcome. For example, when the trial consists of throwing two dice, the set of all outcomes with a sum of 7 pips may constitute an event, whereas outcomes with an odd number of pips may constitute another event. If the outcome is the element of the elementary event of two pips on the first die and five on the second, then both of the events, "7 pips" and "odd number of pips", are said to have happened.
- The probability measure P is a function returning an event's probability. A probability is a real number between zero (impossible events have probability zero, though probability-zero events are not necessarily impossible) and one (the event happens almost surely, with almost total certainty). Thus P is a function $\mathrm{P}:\mathcal{F}\to[0,1]$. The probability measure function must satisfy two simple requirements: First, the probability of a countable union of mutually exclusive events must be equal to the countable sum of the probabilities of each of these events. For example, the probability of the union of the mutually exclusive events Head and Tail in the random experiment of one coin toss, $\mathrm{P}(\mathrm{Head}\cup\mathrm{Tail})$, is the sum of probability for Head and the probability for Tail, $\mathrm{P}(\mathrm{Head})+\mathrm{P}(\mathrm{Tail})$. Second, the probability of the sample space Ω must be equal to 1 (which accounts for the fact that, given an execution of the model, some outcome must occur). In the previous example the probability of the set of outcomes $\mathrm{P}(\{\mathrm{Head},\mathrm{Tail}\})$ must be equal to one, because it is entirely certain that the outcome will be either Head or Tail (the model neglects any other possibility) in a single coin toss.

Not every subset of the sample space Ω must necessarily be considered an event: some of the subsets are simply not of interest, others cannot be "measured". This is not so obvious in a case like a coin toss. In a different example, one could consider javelin throw lengths, where the events typically are intervals like "between 60 and 65 meters" and unions of such intervals, but not sets like the "irrational numbers between 60 and 65 meters".

In short, a probability space is a measure space such that the measure of the whole space is equal to one.

The expanded definition is the following: a probability space is a triple (, ,) consisting of:

- The sample space Ω — an arbitrary non-empty set.

- The σ-algebra $\mathcal{F} \subseteq 2^{\Omega}$ (also called σ-field) — a set of subsets of Ω, called events, such that:
  - $\mathcal{F}$ contains the sample space: $\Omega \in \mathcal{F}$,
  - $\mathcal{F}$ is closed under complements: if $A \in \mathcal{F}$, then also $(\Omega \setminus A) \in \mathcal{F}$,
  - $\mathcal{F}$ is closed under countable unions: if $A_i \in \mathcal{F}$ for $i = 1,2,\ldots$, then also $\left(\bigcup_{i=1}^{\infty} A_i\right) \in \mathcal{F}$.
    - The corollary from the previous two properties and De Morgan's law is that $\mathcal{F}$ is also closed under countable intersections: if $A_i \in \mathcal{F}$ for $i = 1,2,\ldots$, then also $\left(\bigcap_{i=1}^{\infty} A_i\right) \in \mathcal{F}$.
- The probability measure $P : \mathcal{F} \to [0,1]$ — a function on $\mathcal{F}$ such that:
  - P is countably additive (also called σ-additive): if $\{A_i\}_{i=1}^{\infty} \subseteq \mathcal{F}$ is a countable collection of pairwise disjoint sets, then $P(\bigcup_{i=1}^{\infty} A_i) = \sum_{i=1}^{\infty} P(A_i)$,
  - The measure of entire sample space is equal to one: $P(\Omega) = 1$.

Discrete Case

Discrete probability theory needs only at most countable sample spaces Ω. Probabilities can be ascribed to points of Ω by the probability mass function $p : \Omega \to [0,1]$ such that $\sum_{\omega \in \Omega} p(\omega) = 1$. All subsets of Ω can be treated as events (thus, $\mathcal{F} = 2^{\Omega}$ is the power set). The probability measure takes the simple form:

$$(*) \qquad P(A) = \sum_{\omega \in A} p(\omega) \quad \text{for all } A \subseteq \Omega.$$

The greatest σ-algebra $\mathcal{F} = 2^{\Omega}$ describes the complete information. In general, a σ-algebra $\mathcal{F} \subseteq 2^{\Omega}$ corresponds to a finite or countable partition $\Omega = B_1 \cup B_2 \cup \ldots$, the general form of an event $A \in \mathcal{F}$ being $A = B_{k_1} \cup B_{k_2} \cup \ldots$.

The case $p(\omega) = 0$ is permitted by the definition, but rarely used, since such ω can safely be excluded from the sample space.

General Case

If Ω is uncountable, still, it may happen that $p(\omega) \neq 0$ for some ω; such ω are called atoms. They are an at most countable (maybe empty) set, whose probability is the sum of probabilities of all atoms. If this sum is equal to 1 then all other points can safely be excluded from the sample space, returning us to the discrete case. Otherwise, if the sum of probabilities of all atoms is between 0 and 1, then the probability space decomposes into a discrete (atomic) part (maybe empty) and a non-atomic part.

Non-atomic Case

If $p(\omega) = 0$ for all $\omega \in \Omega$ (in this case, Ω must be uncountable, because otherwise $P(\Omega)=1$ could not be satisfied), then equation (∗) fails: the probability of a set is not necessarily the sum over the probabilities of its elements, as summation is only defined for countable numbers of elements. This makes the probability space theory much more technical. A formulation stronger than summation, measure theory is applicable. Initially the probabilities are ascribed to some "generator" sets. Then a limiting procedure allows assigning probabilities to sets that are limits of sequences of generator sets, or limits of limits, and so on. All these sets are the σ-algebra $\mathcal{F}$. Sets belonging to $\mathcal{F}$ are called measurable. In general they are much more complicated than generator sets, but much better than non-measurable sets.

Complete Probability Space

A probability space $(\Omega, \mathcal{F}, P)$ is said to be a complete probability space if for all $B \in \mathcal{F}$ with $P(B)=0$ and all $A \subset B$ one has $A \in \mathcal{F}$. Often, the study of probability spaces is restricted to complete probability spaces.

Discrete Examples

Example: If the experiment consists of just one flip of a fair coin, then the outcome is either heads or tails: $\Omega = \{H, T\}$. The σ-algebra $\mathcal{F} = 2^{\Omega}$ contains $2^2 = 4$ events, namely: $\{H\}$ ("heads"), ("tails"), $\{T\}$ ("neither heads nor tails"), and $\{\}$ ("either heads or tails"); in other words, $\mathcal{F} = \{\{\}, \{H\}, \{T\}, \{H, T\}\}$. There is a fifty percent chance of tossing heads and fifty percent for tails, so the probability measure in this example is , $P(\{\}) = 0$, $P(\{H\}) = 0.5$,, $P(\{T\}) = 0.5$,.

Example: The fair coin is tossed three times. There are 8 possible outcomes: Ω = {HHH, HHT, HTH, HTT, THH, THT, TTH, TTT} (here "HTH" for example means that first time the coin landed heads, the second time tails, and the last time heads again). The complete information is described by the σ-algebra $\mathcal{F} = 2^{\Omega}$ of $2^8 = 256$ events, where each of the events is a subset of Ω.

Alice knows the outcome of the second toss only. Thus her incomplete information is described by the partition $\Omega = A_1 \sqcup A_2$ = {HHH, HHT, THH, THT} ⊔ {HTH, HTT, TTH, TTT}, where ⊔ is the *disjoint union*, and the corresponding σ-algebra $\mathcal{F}_{Alice} = \{\{\}, A_1, A_2, \Omega\}$. Bryan knows only the total number of tails. His partition contains four parts: $\Omega = B_0 \sqcup B_1 \sqcup B_2 \sqcup B_3$ = {HHH} ⊔ {HHT, HTH, THH} ⊔ {TTH, THT, HTT} ⊔ {TTT}; accordingly, his σ-algebra $\mathcal{F}_{Bryan}$ contains $2^4 = 16$ events.

The two σ-algebras are incomparable: neither $\mathcal{F}_{Alice} \subseteq \mathcal{F}_{Bryan}$ nor $\mathcal{F}_{Bryan} \subseteq \mathcal{F}_{Alice}$; both are sub-σ-algebras of 2^{Ω}.

Example: If 100 voters are to be drawn randomly from among all voters in California and asked whom they will vote for governor, then the set of all sequences of 100

Californian voters would be the sample space Ω. We assume that sampling without replacement is used: only sequences of 100 *different* voters are allowed. For simplicity an ordered sample is considered, that is a sequence {Alice, Bryan} is different from {Bryan, Alice}. We also take for granted that each potential voter knows exactly his/her future choice, that is he/she doesn't choose randomly.

Alice knows only whether or not Arnold Schwarzenegger has received at least 60 votes. Her incomplete information is described by the σ-algebra $\mathcal{F}_{\text{Alice}}$ that contains: (1) the set of all sequences in Ω where at least 60 people vote for Schwarzenegger; (2) the set of all sequences where fewer than 60 vote for Schwarzenegger; (3) the whole sample space Ω; and (4) the empty set $\emptyset$.

Bryan knows the exact number of voters who are going to vote for Schwarzenegger. His incomplete information is described by the corresponding partition $\Omega = B_0 \sqcup B_1 \ldots \sqcup B_{100}$ and the σ-algebra $\mathcal{F}_{\text{Bryan}}$ consists of 2^{101} events.

In this case Alice's σ-algebra is a subset of Bryan's: $\mathcal{F}_{\text{Alice}} \subset \mathcal{F}_{\text{Bryan}}$. Bryan's σ-algebra is in turn a subset of the much larger "complete information" σ-algebra 2^{Ω} consisting of $2^{n(n-1)\ldots(n-99)}$ events, where n is the number of all potential voters in California.

Non-atomic Examples

Example: A number between 0 and 1 is chosen at random, uniformly. Here $\Omega = [0,1]$, $\mathcal{F}$ is the σ-algebra of Borel sets on Ω, and P is the Lebesgue measure on [0,1].

In this case the open intervals of the form (a,b), where $0 < a < b < 1$, could be taken as the generator sets. Each such set can be ascribed the probability of $P((a,b)) = (b - a)$, which generates the Lebesgue measure on [0,1], and the Borel σ-algebra on Ω.

Example: A fair coin is tossed endlessly. Here one can take $\Omega = \{0,1\}^{\infty}$, the set of all infinite sequences of numbers 0 and 1. Cylinder sets $\{(x_1, x_2, \ldots) \in \Omega : x_1 = a_1, \ldots, x_n = a_n\}$ may be used as the generator sets. Each such set describes an event in which the first n tosses have resulted in a fixed sequence $(a_1, \ldots, a_n)$, and the rest of the sequence may be arbitrary. Each such event can be naturally given the probability of 2^{-n}.

These two non-atomic examples are closely related: a sequence $(x_1,x_2,\ldots) \in \{0,1\}^{\infty}$ leads to the number $2^{-1}x_1 + 2^{-2}x_2 + \ldots \in [0,1]$. This is not a one-to-one correspondence between $\{0,1\}^{\infty}$ and [0,1] however: it is an isomorphism modulo zero, which allows for treating the two probability spaces as two forms of the same probability space. In fact, all non-pathological non-atomic probability spaces are the same in this sense. They are so-called standard probability spaces. Basic applications of probability spaces are insensitive to standardness. However, non-discrete conditioning is easy and natural on standard probability spaces, otherwise it becomes obscure.

Probability Distribution

Any probability distribution defines a probability measure.

Random Variables

A random variable X is a measurable function $X: \Omega \rightarrow S$ from the sample space Ω to another measurable space S called the *state space*. If $A \subset S$, the notation $\Pr(X \in A)$ is a commonly used shorthand for $P(\{\omega \in \Omega: X(\omega) \in A\})$.

Defining the Events in Terms of the Sample Space

If Ω is countable we almost always define $\mathcal{F}$ as the power set of Ω, i.e. $\mathcal{F} = 2^{\Omega}$ which is trivially a σ-algebra and the biggest one we can create using Ω. We can therefore omit and just write (Ω,P) to define the probability space.

On the other hand, if Ω is uncountable and we use $\mathcal{F} = 2^{\Omega}$ we get into trouble defining our probability measure P because $\mathcal{F}$ is too "large", i.e. there will often be sets to which it will be impossible to assign a unique measure. In this case, we have to use a smaller σ-algebra $\mathcal{F}$, for example the Borel algebra of Ω, which is the smallest σ-algebra that makes all open sets measurable.

Conditional Probability

Kolmogorov's definition of probability spaces gives rise to the natural concept of conditional probability. Every set A with non-zero probability (that is, $P(A) > 0$) defines another probability measure,

$$P(B \mid A) = \frac{P(B \cap A)}{P(A)}$$

on the space. This is usually pronounced as the "probability of B given A".

For any event B such that $P(B) > 0$ the function Q defined by $Q(A) = P(A|B)$ for all events A is itself a probability measure.

Independence

Two events, A and B are said to be independent if $P(A \cap B)=P(A)P(B)$. Two random variables, X and Y, are said to be independent if any event defined in terms of X is independent of any event defined in terms of Y. Formally, they generate independent σ-algebras, where two σ-algebras G and H, which are subsets of F are said to be independent if any element of G is independent of any element of H.

Mutual Exclusivity

Two events, A and B are said to be mutually exclusive or *disjoint* if the occurrence of one implies the non-occurrence of the other, i.e., their intersection is empty. This is a stronger condition than the probability of their intersection being zero.

If A and B are disjoint events, then $P(A \cup B) = P(A) + P(B)$. This extends to a (finite or countably infinite) sequence of events. However, the probability of the union of an uncountable set of events is not the sum of their probabilities. For example, if Z is a normally distributed random variable, then $P(Z=x)$ is 0 for any x, but $P(Z \in \mathbf{R}) = 1$. The event $A \cap B$ is referred to as "A and B", and the event $A \cup B$ as "A or B".

Sample Space

In probability theory, the sample space (also called sample description space or possibility space) of an experiment or random trial is the set of all possible outcomes or results of that experiment. A sample space is usually denoted using set notation, and the possible ordered outcomes are listed as elements in the set. It is common to refer to a sample space by the labels S, Ω, or U (for "universal set"). The elements of a sample space may be numbers, words, letters, or symbols. They can also be finite, countably infinite, or uncountably infinite.

For example, if the experiment is tossing a coin, the sample space is typically the set {head, tail}, commonly written {H, T}. For tossing two coins, the corresponding sample space would be {(head,head), (head,tail), (tail,head), (tail,tail)}, commonly written {HH, HT, TH, TT}. If the sample space is unordered, it becomes {{head,head}, {head,tail}, {tail,tail}}.

For tossing a single six-sided die, the typical sample space is {1, 2, 3, 4, 5, 6} (in which the result of interest is the number of pips facing up). A subset of the sample space is an event, denoted by E. Referring to the experiment of tossing the coin, the possible events include E={H} and E={T}.

A well-defined sample space is one of three basic elements in a probabilistic model (a probability space); the other two are a well-defined set of possible events (a sigma-algebra) and a probability assigned to each event (a probability measure function).

Another way to look as a sample space is visually. The sample space is typically represented by a rectangle, and the outcomes of the sample space denoted by points within the rectangle. The events are represented by ovals, and the points enclosed within the oval make up the event.

Conditions of a Sample Space

A set Ω with outcomes $s_1, s_2, \ldots, s_n$ (i.e. $s_1, s_2, \ldots, s_n$) must meet some conditions in order to be a sample space:

- The outcomes must be mutually exclusive, i.e. if s_j takes place, then no other s_i will take place, $\forall i, j = 1, 2, \ldots, n \quad i \neq j$.

- The outcomes must be collectively exhaustive, i.e., on every experiment (or random trial) there will always take place some outcome $s_i \in \Omega$ for $i \in \{1,2,\ldots,n\}$.
- The sample space (Ω) must have the right granularity depending on what we are interested in. We must remove irrelevant information from the sample space. In other words, we must choose the right abstraction (forget some irrelevant information).

For instance, in the trial of tossing a coin, we could have as a sample space $\Omega_1 = \{H, T\}$, where H stands for *heads* and T for *tails*. Another possible sample space could be $\Omega_2 = \{H\&R, H\&NR, TR, T\&NR\}$. Here, R stands for *rains* and NR *not rains*. Obviously, Ω_1 is a better choice than Ω_2 as we do not care about how the weather affects the tossing of a coin.

Multiple Sample Spaces

For many experiments, there may be more than one plausible sample space available, depending on what result is of interest to the experimenter. For example, when drawing a card from a standard deck of fifty-two playing cards, one possibility for the sample space could be the various ranks (Ace through King), while another could be the suits (clubs, diamonds, hearts, or spades). A more complete description of outcomes, however, could specify both the denomination and the suit, and a sample space describing each individual card can be constructed as the Cartesian product of the two sample spaces noted above (this space would contain fifty-two equally likely outcomes). Still other sample spaces are possible, such as {right-side up, up-side down} if some cards have been flipped when shuffling.

Equally Likely Outcomes

Flipping a coin leads to a sample space composed of two outcomes that are almost equally likely.

Up or down? Flipping a brass tack leads to a sample space composed of two outcomes that are not equally likely.

Some treatments of probability assume that the various outcomes of an experiment are always defined so as to be equally likely. For any sample space with N equally likely outcomes, each outcome is assigned the probability 1/N. However, there are experiments that are not easily described by a sample space of equally likely outcomes—for example, if one were to toss a thumb tack many times and observe whether it landed with its point upward or downward, there is no symmetry to suggest that the two outcomes should be equally likely.

Though most random phenomena do not have equally likely outcomes, it can be helpful to define a sample space in such a way that outcomes are at least approximately equally likely, since this condition significantly simplifies the computation of probabilities for events within the sample space. If each individual outcome occurs with the same probability, then the probability of any event becomes simply:

$$P(\text{event}) = \frac{\text{number of outcomes in event}}{\text{number of outcomes in sample space}}$$

For example, if two dice are thrown to generate two uniformly distributed integers, D_1 and D_2, each in the range [1...6], the 36 ordered pairs (D_1 , D_2) constitute a sample space of equally likely events. In this case, the above formula applies, such that the probability of a certain sum, say $D_1 + D_2 = 5$ is easily shown to be 4/36, since 4 of the 36 outcomes produce 5 as a sum. On the other hand, the sample space of the 11 possible sums, {2, ...,12} are not equally likely outcomes, so the formula would give an incorrect result (1/11).

Another example is having four pens in a bag. One pen is red, one is green, one is blue, and one is purple. Each pen has the same chance of being taken out of the bag. The sample space S={red, green, blue, purple}, consists of equally likely events. Here, P(red)=P(blue)=P(green)=P(purple)=1/4.

Simple Random Sample

In statistics, inferences are made about characteristics of a population by studying a sample of that population's individuals. In order to arrive at a sample that presents an unbiased estimate of the true characteristics of the population, statisticians often seek to study a simple random sample—that is, a sample in which every individual in the population is equally likely to be included. The result of this is that every possible combination of individuals who could be chosen for the sample has an equal chance to be the sample that is selected (that is, the space of simple random samples of a given size from a given population is composed of equally likely outcomes).

Infinitely Large Sample Spaces

In an elementary approach to probability, any subset of the sample space is usually called an event. However, this gives rise to problems when the sample space is continuous, so that a more precise definition of an event is necessary. Under this definition only measurable subsets of the sample space, constituting a σ-algebra over the sample space itself, are considered events.

An example of an infinitely large sample space is measuring the lifetime of a light bulb. The corresponding sample space would be [0, infinity).

Event in Probability Theory

In probability theory, an event is a set of outcomes of an experiment (a subset of the sample space) to which a probability is assigned. A single outcome may be an element of many different events, and different events in an experiment are usually not equally likely, since they may include very different groups of outcomes. An event defines a complementary event, namely the complementary set (the event *not* occurring), and together these define a Bernoulli trial: did the event occur or not?

Typically, when the sample space is finite, any subset of the sample space is an event (*i.e.* all elements of the power set of the sample space are defined as events). However, this approach does not work well in cases where the sample space is uncountably infinite. So, when defining a probability space it is possible, and often necessary, to exclude certain subsets of the sample space from being events.

Example:

If we assemble a deck of 52 playing cards with no jokers, and draw a single card from the deck, then the sample space is a 52-element set, as each card is a possible outcome. An event, however, is any subset of the sample space, including any singleton set (an elementary event), the empty set (an impossible event, with probability zero) and the sample space itself (a certain event, with probability one). Other events are proper subsets of the sample space that contain multiple elements. So, for example, potential events include:

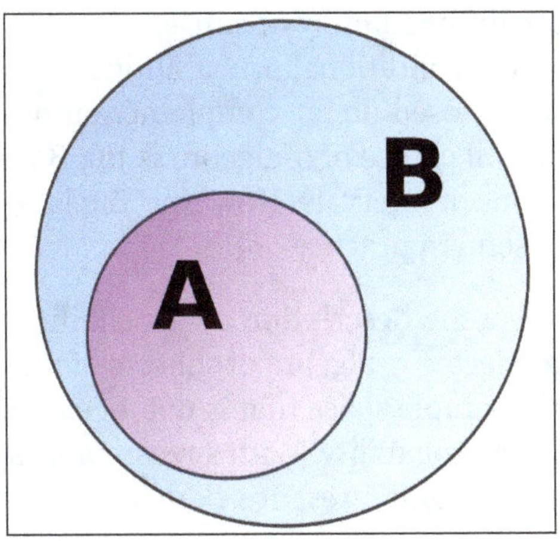

An Euler diagram of an event. *B* is the sample space and *A* is an event. By the ratio of their areas, the probability of *A* is approximately 0.4.

- "Red and black at the same time without being a joker" (0 elements),
- "The 5 of Hearts" (1 element),
- "A King" (4 elements),
- "A Face card" (12 elements),
- "A Spade" (13 elements),
- "A Face card or a red suit" (32 elements),
- "A card" (52 elements).

Since all events are sets, they are usually written as sets (e.g. {1, 2, 3}), and represented graphically using Venn diagrams. In the situation where each outcome in the sample space Ω is equally likely, the probability P of an event *A* is the following formula:

$$P(A) = \frac{|A|}{|\Omega|}\left(\text{alternatively}: \Pr(A) = \frac{|A|}{|\Omega|}\right)$$

This rule can readily be applied to each of the example events above.

Events in Probability Spaces

Defining all subsets of the sample space as events works well when there are only finitely many outcomes, but gives rise to problems when the sample space is infinite. For many standard probability distributions, such as the normal distribution, the sample space is the set of real numbers or some subset of the real numbers. Attempts to define probabilities for all subsets of the real numbers run into difficulties when one considers 'badly behaved' sets, such as those that are nonmeasurable. Hence, it is necessary to restrict attention to a more limited family of subsets. For the standard tools of probability theory, such as joint and conditional probabilities, to work, it is necessary to use a σ-algebra, that is, a family closed under complementation and countable unions of its members. The most natural choice of σ-algebra is the Borel measurable set derived from unions and intersections of intervals. However, the larger class of Lebesgue measurable sets proves more useful in practice.

In the general measure-theoretic description of probability spaces, an event may be defined as an element of a selected σ-algebra of subsets of the sample space. Under this definition, any subset of the sample space that is not an element of the σ-algebra is not an event, and does not have a probability. With a reasonable specification of the probability space, however, all *events of interest* are elements of the σ-algebra.

Even though events are subsets of some sample space Ω, they are often written as predicates or indicators involving random variables. For example, if *X* is a real-valued random variable defined on the sample space Ω, the event,

$$\{\omega \in \Omega \mid u < X(\omega) \leq v\}$$

can be written more conveniently as, simply,

$$u < X \leq v.$$

This is especially common in formulas for a probability, such as,

$$\Pr(u < X \leq v) = F(v) - F(u).$$

The set $u < X \leq v$ is an example of an inverse image under the mapping *X* because $\omega \in X^{-1}((u,v])$ if and only if $u < X(\omega) \leq v$.

Elementary Event

In probability theory, an elementary event (also called an atomic event or sample point) is an event which contains only a single outcome in the sample space. Using set theory terminology, an elementary event is a singleton. Elementary events and their corresponding outcomes are often written interchangeably for simplicity, as such an event corresponds to precisely one outcome.

The following are examples of elementary events:

- All sets {k}, where k ∈ N if objects are being counted and the sample space is S = {0, 1, 2, 3, ...} (the natural numbers).
- {HH}, {HT}, {TH} and {TT} if a coin is tossed twice. S = {HH, HT, TH, TT}. H stands for heads and T for tails.
- All sets {x}, where x is a real number. Here X is a random variable with a normal distribution and S = (−∞, +∞). This example shows that, because the probability of each elementary event is zero, the probabilities assigned to elementary events do not determine a continuous probability distribution.

Probability of an Elementary Event

Elementary events may occur with probabilities that are between zero and one (inclusively). In a discrete probability distribution whose sample space is finite, each elementary event is assigned a particular probability. In contrast, in a continuous distribution, individual elementary events must all have a probability of zero because there are infinitely many of them— then non-zero probabilities can only be assigned to non-elementary events.

Some “mixed” distributions contain both stretches of continuous elementary events and some discrete elementary events; the discrete elementary events in such distributions can be called atoms or atomic events and can have non-zero probabilities.

Under the measure-theoretic definition of a probability space, the probability of an elementary event need not even be defined. In particular, the set of events on which probability is defined may be some σ-algebra on S and not necessarily the full power set.

Complementary Event

In probability theory, the complement of any event A is the event [not A], i.e. the event that A does not occur. The event A and its complement [not A] are mutually exclusive and exhaustive. Generally, there is only one event B such that A and B are both mutually exclusive and exhaustive; that event is the complement of A. The complement of an event A is usually denoted as A', A^c, $\neg A$ or A. Given an event, the event and its complementary event define a Bernoulli trial: did the event occur or not?

For example, if a typical coin is tossed and one assumes that it cannot land on its edge, then it can either land showing “heads” or “tails.” Because these two outcomes are mutually exclusive (i.e. the coin cannot simultaneously show both heads and tails) and collectively exhaustive (i.e. there are no other possible outcomes not represented between these two), they are therefore each other’s complements. This means that [heads] is logically equivalent to [not tails], and [tails] is equivalent to [not heads].

Complement Rule

In a random experiment, the probabilities of all possible events (the sample space) must total to 1— that is, some outcome must occur on every trial. For two events to be complements, they must be collectively exhaustive, together filling the entire sample space. Therefore, the probability of an event’s complement must be unity minus the probability of the event. That is, for an event A,

$$P(A^c) = 1 - P(A).$$

Equivalently, the probabilities of an event and its complement must always total to 1. This does not, however, mean that any two events whose probabilities total to 1 are each other’s complements; complementary events must also fulfill the condition of mutual exclusivity.

Example of the Utility of this Concept

Suppose one throws an ordinary six-sided die eight times. What is the probability that one sees a “1” at least once?

It may be tempting to say that:

Pr(["1" on 1st trial] or ["1" on second trial] or ... or ["1" on 8th trial])

= Pr("1" on 1st trial) + Pr("1" on second trial) + ... + P("1" on 8th trial)

= 1/6 + 1/6 + ... + 1/6.

= 8/6 = 1.3333... (...and this is clearly wrong.)

That cannot be right because a probability cannot be more than 1. The technique is wrong because the eight events whose probabilities got added are not mutually exclusive.

One may resolve this overlap by the principle of inclusion-exclusion, or in this case one may instead more simply find the probability of the complementary event and subtract it from 1, thus:

Pr(at least one "1") = 1 – Pr(no "1"s)

= 1 – Pr([no "1" on 1st trial] and [no "1" on 2nd trial] and ... and [no "1" on 8th trial])

= 1 – Pr(no "1" on 1st trial) × Pr(no "1" on 2nd trial) × ... × Pr(no "1" on 8th trial)

= 1 –(5/6) × (5/6) × ... × (5/6)

= 1 – $(5/6)^8$

= 0.7674...

Independent Event

This is a fundamental notion in probability theory, as in statistics and the theory of stochastic processes.

Two events are independent, statistically independent, or stochastically independent if the occurrence of one does not affect the probability of occurrence of the other (equivalently, does not affect the odds). Similarly, two random variables are independent if the realization of one does not affect the probability distribution of the other.

When dealing with collections of more than two events, a weak and a strong notion of independence need to be distinguished. The events are called pairwise independent if any two events in the collection are independent of each other, while saying that the events are mutually independent (or collectively independent) intuitively means that each event is independent of any combination of other events in the collection. Similar notions for collections of random variables.

The name "mutual independence" (same as "collective independence") seems the outcome of a pedagogical choice, merely to distinguish the stronger notion from "pairwise independence" which is a weaker notion. In the advanced literature of probability theory, statistics and stochastic processes, the stronger notion is simply named independence with no modifier. It is stronger since independence implies pairwise independence, but not the other way around.

For Events

Two Events

Two events A and B are independent (often written as $A \perp B$ or $A \perp\!\!\!\perp B$) if and only if their joint probability equals the product of their probabilities:

$$P(A \cap B) = P(A)P(B)$$

Why this defines independence is made clear by rewriting with conditional probabilities:

$$P(A \cap B) = P(A)P(B) \Leftrightarrow P(A) = \frac{P(A \cap B)}{P(B)} = P(A \mid B).$$

and similarly, $P(A \cap B) = P(A)P(B) \Leftrightarrow P(B) = P(B \mid A)$.

Thus, the occurrence of B does not affect the probability of A, and vice versa. Although the derived expressions may seem more intuitive, they are not the preferred definition, as the conditional probabilities may be undefined if $P(A)$ or $P(B)$ are 0. Furthermore, the preferred definition makes clear by symmetry that when A is independent of B, B is also independent of A.

Log Probability and Information Content

Stated in terms of log probability, two events are independent if and only if the log probability of the joint event is the sum of the log probability of the individual events:

$$\log P(A \cap B) = \log P(A) + \log P(B)$$

In information theory, negative log probability is interpreted as information content, and thus two events are independent if and only if the information content of the combined event equals the sum of information content of the individual events:

$$I(A \cap B) = I(A) + I(B)$$

Odds

Stated in terms of odds, two events are independent if and only if the odds ratio of A

and B is unity (1). Analogously with probability, this is equivalent to the conditional odds being equal to the unconditional odds:

$$O(A \mid B) = O(A) \text{ and } O(B \mid A) = O(B),$$

or to the odds of one event, given the other event, being the same as the odds of the event, given the other event not occurring:

$$O(A \mid B) = O(A \mid \neg B) \text{ and } O(B \mid A) = O(B \mid \neg A).$$

The odds ratio can be defined as: $O(A \mid B) : O(A \mid \neg B)$,

or symmetrically for odds of B given A, and thus is 1 if and only if the events are independent.

More than Two Events

A finite set of events $\{A_i\}_{i=1}^{n}$ is pairwise independent if every pair of events is independent—that is, if and only if for all distinct pairs of indices m, k,

$$P(A_m \cap A_k) = P(A_m)P(A_k)$$

A finite set of events is mutually independent if every event is independent of any intersection of the other events—that is, if and only if for every $k \le n$ and for every k-element subset of events $\{B_i\}_{i=1}^{k}$ of $\{A_i\}_{i=1}^{n}$,

$$P\left(\bigcap_{i=1}^{k} B_i\right) = \prod_{i=1}^{k} P(B_i)$$

This is called the *multiplication rule* for independent events. Note that it is not a single condition involving only the product of all the probabilities of all single events; it must hold true for all subsets of events.

For more than two events, a mutually independent set of events is (by definition) pairwise independent; but the converse is not necessarily true.

For Real Valued Random Variables

Two Random Variables

Two random variables X and Y are independent if and only if (iff) the elements of the π-system generated by them are independent; that is to say, for every x and y, the events $\{X \le x\}$ and $\{Y \le y\}$ are independent events. That is, X and Y with cumulative distribution functions $F_X(x)$ and $F_Y(y)$,, are independent iff the combined random variable (X, Y) has a joint cumulative distribution function:

$$F_{X,Y}(x, y) = F_X(x)F_Y(y) \quad \text{for all } x, y$$

or equivalently, if the probability densities $f_X(x)$ and $f_Y(y)$ and the joint probability density $f_{X,Y}(x,y)$ exist, $f_{X,Y}(x,y) = f_X(x) f_Y(y) \quad \text{for all } x,y.$

More than Two Random Variables

A finite set of n random variables $\{X_1,\ldots,X_n\}$ is pairwise independent if and only if every pair of random variables is independent. Even if the set of random variables is pairwise independent, it is not necessarily mutually independent as defined next.

A finite set of n random variables $\{X_1,\ldots,X_n\}$ is mutually independent if and only if for any sequence of numbers $\{x_1,\ldots,x_n\}$, the events $\{X_1 \le x_1\},\ldots,\{X_n \le x_n\}$ are mutually independent events. This is equivalent to the following condition on the joint cumulative distribution function $F_{X_1,\ldots,X_n}(x_1,\ldots,x_n)$. A finite set of n random variables $\{X_1,\ldots,X_n\}$ is mutually independent if and only if,

$$F_{X_1,\ldots,X_n}(x_1,\ldots,x_n) = F_{X_1}(x_1)\cdot\ldots\cdot F_{X_n}(x_n) \quad \text{for all } x_1,\ldots,x_n$$

Notice that it is not necessary here to require that the probability distribution factorizes for all possible k – element subsets as in the case for n events. This is not required because e.g. $F_{X_1,X_2,X_3}(x_1,x_2,x_3) = F_{X_1}(x_1)\cdot F_{X_2}(x_2)\cdot F_{X_3}(x_3)$ implies $F_{X_1,X_3}(x_1,x_3) = F_{X_1}(x_1)\cdot F_{X_3}(x_3)$.

The measure-theoretically inclined may prefer to substitute events $\{X \in A\}$ for events $\{X \le x\}$ in the above definition, where A is any Borel set. That definition is exactly equivalent to the one above when the values of the random variables are real numbers. It has the advantage of working also for complex-valued random variables or for random variables taking values in any measurable space (which includes topological spaces endowed by appropriate σ-algebras).

For Real Valued Random Vectors

Two random vectors $\mathbf{X} = (X_1,\ldots,X_m)^T$ and $\mathbf{Y} = (Y_1,\ldots,Y_n)^T$ are called independent if,

$$F_{\mathbf{X},\mathbf{Y}}(\mathbf{x},\mathbf{y}) = F_{\mathbf{X}}(\mathbf{x})\cdot F_{\mathbf{Y}}(\mathbf{y}) \quad \text{for all } \mathbf{x},\mathbf{y}$$

where $F_{\mathbf{X}}(\mathbf{x})$ and $F_{\mathbf{Y}}(\mathbf{y})$ denote the cumulative distribution functions of $\mathbf{X}$ and $\mathbf{Y}$ and $F_{\mathbf{X},\mathbf{Y}}(\mathbf{x},\mathbf{y})$ denotes their joint cumulative distribution function. Independence of $\mathbf{X}$ and $\mathbf{Y}$ is often denoted by $\mathbf{X} \perp\!\!\!\perp \mathbf{Y}$. Written component-wise, $\mathbf{X}$ and $\mathbf{Y}$ are called independent if,

$$F_{\mathbf{X},\mathbf{Y}}(\mathbf{x},\mathbf{y}) = F_{\mathbf{X}}(\mathbf{x})\cdot F_{\mathbf{Y}}(\mathbf{y}) \quad \text{for all } \mathbf{x},\mathbf{y}.$$

For Stochastic Processes

For One Stochastic Process

The definition of independence may be extended from random vectors to a stochastic process. Thereby it is required for an independent stochastic process that the random variables obtained by sampling the process at any n times $t_1,\ldots,t_n$ are independent random variables for any n.

Formally, a stochastic process $\{X_t\}_{t\in\mathcal{T}}$ is called independent, if and only if for all $n\in\mathbb{N}$ and for all $t_1,\ldots,t_n\in\mathcal{T}$,

$$F_{X_{t_1},\ldots,X_{t_n}}(x_1,\ldots,x_n)=F_{X_{t_1}}(x_1)\cdot\ldots\cdot F_{X_{t_n}}(x_n)\quad \text{for all } x_1,\ldots,x_n$$

where $F_{X_{t_1},\ldots,X_{t_n}}(x_1,\ldots,x_n)=P(X(t_1)\le x_1,\ldots,X(t_n)\le x_n)$. Notice that independence of a stochastic process is a property *within* a stochastic process, not between two stochastic processes.

For Two Stochastic Processes

Independence of two stochastic processes is a property between two stochastic processes $\{X_t\}_{t\in\mathcal{T}}$ and $\{Y_t\}_{t\in\mathcal{T}}$ that are defined on the same probability space $(\Omega,\mathcal{F},P)$. Formally, two stochastic processes $\{X_t\}_{t\in\mathcal{T}}$ and $\{Y_t\}_{t\in\mathcal{T}}$ are said to be independent if for all $n\in\mathbb{N}$ and for all $t_1,\ldots,t_n\in\mathcal{T}$, the random vectors $(X(t_1),\ldots,X(t_n))$ and $(Y(t_1),\ldots,Y(t_n))$ are independent, i.e. if,

$$\begin{aligned}&F_{X_{t_1},\ldots,X_{t_n},Y_{t_1},\ldots,Y_{t_n}}(x_1,\ldots,x_n,y_1,\ldots,y_n)\\&=F_{X_{t_1},\ldots,X_{t_n}}(x_1,\ldots,x_n)\cdot F_{Y_{t_1},\ldots,Y_{t_n}}(y_1,\ldots,y_n)\quad \text{for all } x_1,\ldots,x_n\end{aligned}$$

Independent σ-algebras

The definitions above are both generalized by the following definition of independence for σ-algebras. Let $(\Omega,\mathcal{F},P)$. be a probability space and let $\mathcal{A}$ and $\mathcal{B}$ be two sub-σ-algebras of Σ. $\mathcal{A}$ and $\mathcal{B}$ are said to be independent if, whenever $A\in\mathcal{A}$ and $B\in\mathcal{B}$,

$$P(A\cap B)=P(A)P(B).$$

Likewise, a finite family of σ-algebras $(\tau_i)_{i\in I}$, , where I is an index set, is said to be independent if and only if,

$$\forall (A_i)_{i\in I}\in\prod\nolimits_{i\in I}\tau_i : P\left(\bigcap\nolimits_{i\in I}A_i\right)=\prod\nolimits_{i\in I}P(A_i)$$

and an infinite family of σ-algebras is said to be independent if all its finite subfamilies are independent.

The new definition relates to the previous ones very directly:

- Two events are independent (in the old sense) if and only if the σ-algebras that they generate are independent (in the new sense). The σ-algebra generated by an event $E\in\Sigma$ is, by definition,

  $$\sigma(\{E\})=\{\varnothing,E,\Omega\setminus E,\Omega\}.$$

- Two random variables X and Y defined over Ω are independent (in the old sense) if and only if the σ-algebras that they generate are independent (in the new sense). The σ-algebra generated by a random variable Ω taking values in some measurable space S consists, by definition, of all subsets of Ω of the form $X^{-1}(U)$, where U is any measurable subset of S.

Using this definition, it is easy to show that if X and Y are random variables and Y is constant, then X and Y are independent, since the σ-algebra generated by a constant random variable is the trivial σ-algebra $\{\varnothing,\Omega\}$. Probability zero events cannot affect independence so independence also holds if Y is only Pr-almost surely constant.

Properties

Self-independence

Note that an event is independent of itself if and only if,

$$P(A) = P(A \cap A) = P(A) \cdot P(A) \Leftrightarrow P(A) = 0 \text{ or } P(A) = 1.$$

Thus an event is independent of itself if and only if it almost surely occurs or its complement almost surely occurs; this fact is useful when proving zero–one laws.

Expectation and Covariance

If X and Y are independent random variables, then the expectation operator E has the property,

$$E[XY] = E[X]E[Y],$$

and the covariance cov[X, Y] is zero, since we have,

$$\mathrm{cov}[X,Y] = E[XY] - E[X]E[Y].$$

(The converse of these, i.e. the proposition that if two random variables have a covariance of 0 they must be independent, is not true).

Similarly for two stochastic processes $\{X_t\}_{t\in T}$ and $\{Y_t\}_{t\in T}$: If they are independent, then they are uncorrelated.

Characteristic Function

Two random variables X and Y are independent if and only if the characteristic function of the random vector (X, Y) satisfies,

$$\varphi_{(X,Y)}(t,s) = \varphi_X(t) \cdot \varphi_Y(s).$$

In particular the characteristic function of their sum is the product of their marginal, characteristic functions:

$$\varphi_{X+Y}(t) = \varphi_X(t) \cdot \varphi_Y(t),$$

though the reverse implication is not true. Random variables that satisfy the latter condition are called subindependent.

Rolling Dice

The event of getting a 6 the first time a die is rolled and the event of getting a 6 the second time are *independent*. By contrast, the event of getting a 6 the first time a die is rolled and the event that the sum of the numbers seen on the first and second trial is 8 are *not* independent.

Drawing Cards

If two cards are drawn *with* replacement from a deck of cards, the event of drawing a red card on the first trial and that of drawing a red card on the second trial are *independent*. By contrast, if two cards are drawn *without* replacement from a deck of cards, the event of drawing a red card on the first trial and that of drawing a red card on the second trial are *not* independent, because a deck that has had a red card removed has proportionately fewer red cards.

Pairwise and Mutual Independence

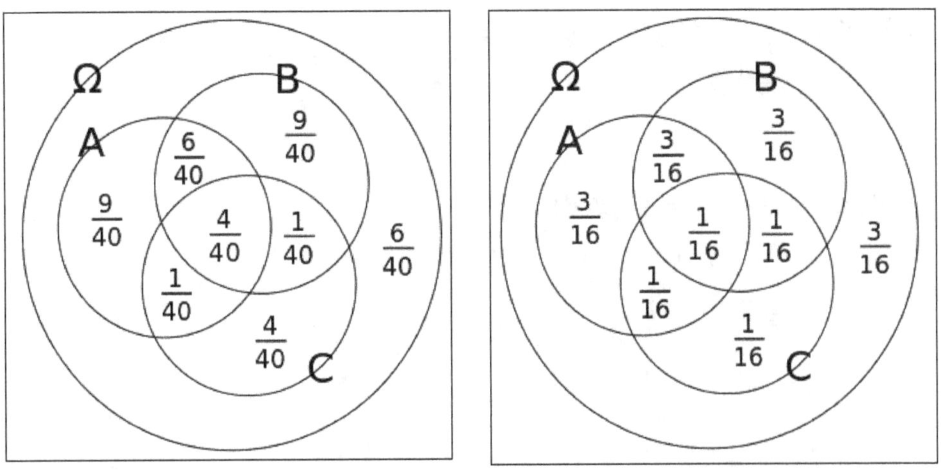

Pairwise independent, but not mutually independent, events.

Mutually independent events.

Consider the two probability spaces shown. In both cases, $P(A) = P(B) = 1/2$ and $P(C) = 1/4$. The random variables in the first space are pairwise independent because $P(A|B)=P(A|C)=1/2=P(A)$, $P(B|A)=P(B|C)=1/2=P(B)$, and $P(C|A)=P(C|B)=1/4=P(C)$; but the three random variables are not mutually independent. The random variables in the second space are both pairwise independent and mutually independent. To illustrate the difference, consider conditioning on two events. In the pairwise independent case, although any one event is independent of each of the other two individually, it is not independent of the intersection of the other two:

$$P(A|BC)=\frac{\frac{4}{40}}{\frac{4}{40}+\frac{1}{40}}=\frac{4}{5}\neq P(A)$$

$$P(B|AC)=\frac{\frac{4}{40}}{\frac{4}{40}+\frac{1}{40}}=\frac{4}{5}\neq P(B)$$

$$P(C|AB)=\frac{\frac{4}{40}}{\frac{4}{40}+\frac{6}{40}}=\frac{2}{5}\neq P(C)$$

In the mutually independent case, however,

$$(A|BC)=\frac{\frac{1}{16}}{\frac{1}{16}+\frac{1}{16}}=\frac{1}{2}=P(A)$$

$$P(B|AC)=\frac{\frac{1}{16}}{\frac{1}{16}+\frac{1}{16}}=\frac{1}{2}=P(B)$$

$$P(C|AB)=\frac{\frac{1}{16}}{\frac{1}{16}+\frac{3}{16}}=\frac{1}{4}=P(C)$$

Mutual Independence

It is possible to create a three-event example in which:

$$P(A\cap B\cap C)=P(A)P(B)P(C),$$

and yet no two of the three events are pairwise independent (and hence the set of events are not mutually independent). This example shows that mutual independence involves requirements on the products of probabilities of all combinations of events, not just the single events as in this example. For another example, take A to be empty and B and C to be identical events with non-zero probability. Then, since B and C are the same event, they are not independent, but the probability of the intersection of the events is zero, the product of the probabilities.

Conditional Independence

For Events

The events A and B are conditionally independent given an event C when,

$$P(A \cap B \mid C) = P(A \mid C) \cdot P(B \mid C).$$

For Random Variables

Intuitively, two random variables X and Y are conditionally independent given Z if, once Z is known, the value of Y does not add any additional information about X. For instance, two measurements X and Y of the same underlying quantity Z are not independent, but they are conditionally independent given Z (unless the errors in the two measurements are somehow connected).

The formal definition of conditional independence is based on the idea of conditional distributions. If X, Y, and Z are discrete random variables, then we define X and Y to be *conditionally independent given* Z if,

$$P(X \le x, Y \le y \mid Z = z) = P(X \le x \mid Z = z) \cdot P(Y \le y \mid Z = z)$$

for all x, y and z such that $P(Z = z) > 0$. On the other hand, if the random variables are continuous and have a joint probability density function $f_{XYZ}(x,y,z)$,, then X and Y are conditionally independent given Z if,

$$f_{XY|Z}(x,y \mid z) = f_{X|Z}(x \mid z) \cdot f_{Y|Z}(y \mid z)$$

for all real numbers x, y and z such that $f_Z(z) > 0$.

If discrete X and Y are conditionally independent given Z, then,

$$P(X = x \mid Y = y, Z = z) = P(X = x \mid Z = z)$$

for any x, and z with $P(Z = z) > 0$. That is, the conditional distribution for X given Y and Z is the same as that given Z alone. A similar equation holds for the conditional probability density functions in the continuous case.

Independence can be seen as a special kind of conditional independence, since probability can be seen as a kind of conditional probability given no events.

Mutual Exclusivity

Two events are said to be mutually exclusive events when both cannot occur at the same time. Mutually exclusive events always have a different outcome. Such events are so that when one happens it prevents the second from happening. For example, if the coin toss gives you a "Head" it won't give you a "Tail". These are mutually exclusive events.

Now you might say that any two events are mutually exclusive then Not exactly. Consider two coins that we toss at the same time. The occurrence of either a Head or a Tail on one of these doesn't affect the probability of the occurrence of H or T of the other coin.

There are other kinds of events also. For example, consider a coin that has a Head on both sides or a Tail on both sides. No matter how many times you flip it, it will always be Head (for the first coin) and Tail (for the second coin). So how will the sample space of such an experiment look? It will be either {H} for the first coin and {T} for the second one. Such events have only one point in the sample space and are known as the "Simple Events". Two simple events are always mutually exclusive.

Die or Dicc

A die or dice is essentially a cube of six faces. In the theory of probability, the concept of a die is used to study events and their interrelations. The die has six faces that are different from each other. The probability of getting one of the faces in an event (say a 6 or a 2) is equal to 1/6.

Note: The number of favourable events is 1 because the desirable face is just one. On the other hand, the total number of events is 6 as there are 6 faces of the die. We can say the following about the mutually exclusive events:

If A and B are two sample spaces of their respective events, such that (A ∩ B) = Ø (phi or the empty set represented by 'Ø' contains no element).

Then, P (A ∩ B) = 0 or the probability of A and B happening together is = 0. In other words, the events are mutually exclusive. The symbol ∩ represents intersection or the word 'and'. Hence, P (A ∩ B) means the probability of the occurrence of A and B. It will be zero only if the events are mutually exclusive.

We can also see the probability that either A or B will happen. For example, what is the probability that in a coin toss either a head or a tail will turn up? Well, it is going to happen with absolute certainty. In the words of probability, we can say that this probability = 1.

If A and B are two sample spaces of their respective events, such that $(A \cap B) = \emptyset$ [phi or the empty set represented by '$\emptyset$' contains no element]. Then the probability of either A or B happening will be written as follows:

$P(A \cup B) = P(A) + P(B)$; where the symbol '$\cup$' represents union or the word 'or'. So the probability of occurrence of either A or B when A and B are mutually exclusive events is equal to the probability of occurrence of A plus the probability of occurrence of B.

Example: Three coins are tossed simultaneously. We represent P as the event of getting at least 2 heads. Similarly, Q represents the event of getting no heads and R is the event of getting heads on the second coin. Which of these is mutually exclusive?

Answer: To make it an easy problem, let us build the sample space of each event. For the event 'P' we want to get at least two head. That means we will include all the events that have two or more heads.

In other words, we can write P = {HHT, HTH, THH, HHH}. This set has 4 elementsor events in it i.e. n(P) = 4.

Similarly for the event Q, we can write the sample as Q = {TTT} and n(Q) = 1.

Therefore using similar logic, we can write R = {THT, HHH, HHT, THH} and n(R) = 4

So Q & R and P & R are mutually exclusive as they have nothing in their intersection.

Note that neither P nor Q or R is the sample space here. They are subsets of the sample space. The sample space will have 20 elements in it.

Example: In the above example, what is the probability of getting exactly two heads?

Answer: The set of getting exactly two heads can be written as {HHT, HTH, THH}. This means that there are three favourable outcomes out of a possible 20 outcomes. So the probability will be 3/20.

Boole's Inequality

The other most commonly used name for Boole's inequality is the "union bound". The Boole's inequality is stating that if we have been given some countable events, that is, a finite set of events, then the probability of occurrence of at least one of the events is clearly less than or equal to the sum of the probabilities of the occurrences of these individual events. This can be rewritten mathematically in the following form. Let E_k be a set of events and the probability that E_k is true is $P(E_k)$. The probability that at least any one of $E_1,...,E_n$ is true, is denoted by $P(\cup(i=1)kE_k)$. Then Boole's inequality is given by the following relation:

$$P(\cup_{k=1}^{n} E_k) \leq \Sigma_{k=1}^{n} P(E_k)$$

Where E_k, for k = 1, ...,n are the given set of finite or countable events.

Proof: We will prove the Boole's inequality by using the method of induction: When n = 1, the inequality is, $P(E_1) \leq P(E_1)$ which is true always.

Assume that it also holds for i = k. That is $P(\cup_{c=1}^{k} E_c) \leq \Sigma_{c=1}^{k} P(E_c)$ is true. Now, we will prove that the inequality holds for i = k + 1.

When i = k + 1 we have,

$$P(\cup_{j=1}^{k+1} E_j)$$
$$= P(\cup_{j=1}^{k} E_j \cup E_{k+1})$$
$$= P\left(\cup_{j=1}^{k} E_j\right) + P\left(E_{k+1}\right) - P\left(\cup_{j=1}^{k} E_j \cap E_{k+1}\right)$$

We know that $P\left(\cup_{j=1}^{k} E_j \cap E_{k+1}\right) \geq 0$

$$\Rightarrow P\left(\cup_{j=1}^{k+1} E_j\right) \leq P\left(\cup_{j=1}^{k} E_j\right) + P\left(E_{k+1}\right)$$
$$\Rightarrow P\left(\cup_{j=1}^{k+1} E_j\right) \leq \Sigma_{j=1}^{k} P\left(E_j\right) + P\left(E_{k+1}\right)$$
$$\Rightarrow P\left(\cup_{j=1}^{k+1} E_j\right) \leq \Sigma_{j=1}^{k+1} P\left(E_j\right)$$

We conclude that since the inequality holds for i = k +1 when it holds for i = k, so by the principle of mathematical induction, we can say that $P\left(\cup_{j=1}^{n} E_j\right) \leq \Sigma_{j=1}^{n} P\left(E_j\right)$ for all 'n'.

We can generalize the Boole's inequality in case of probability of events in finite union in order to find their upper and lower bounds. We call these bounds Bonferroni inequalities. For this we are first defining three different summations:

- $X_1 = \Sigma_{j=1}^{n} P(B_i)$
- $X_2 = \Sigma_{1 \leq i < j \leq n} P(B_i, \cap B_j)$
- $X_k = \Sigma_{1 \leq i1 < .... < ik \leq n} P(B_{i1} \cap .... \cap B_{ik})$

For all natural numbers starting from 3 to n. With these three definitions, we move ahead by following observations: For any odd value of k in the set {1, ..., n} we will have, $P(\cup_{j=1}^{n} B_i) \leq \Sigma_{j=1}^{k} -1^{j-1} X_j$ And for any even value of k in the set {2, 4, ..., n} we will have, $P(\cup_{j=1}^{n} B_i) >= \Sigma_{j=1}^{k} -1^{j-1} X_j$.

We can recover the original Boole's inequality when k = 1. When k = n, then we will get the equality that holds true and then the identity that is obtained is known as the inclusion-exclusion principle.

References

- Yates, Daniel S.; Moore, David S; Starnes, Daren S. (2003). The Practice of Statistics (2nd ed.). New York: Freeman. ISBN 978-0-7167-4773-4. Archived from the original on 2005-02-09. Retrieved 2013-07-18
- What-are-probability-axioms-3126567: thoughtco.com, Retrieved 1 June, 2020
- Gallager, Robert G. (2013). Stochastic Processes Theory for Applications. Cambridge University Press. ISBN 978-1-107-03975-9
- Mutually-exclusive-events, probability, quantitative-aptitude: toppr.com, Retrieved 1 May, 2020
- Forbes, Catherine; Evans, Merran; Hastings, Nicholas; Peacock, Brian (2011). Statistical Distributions (4th ed.). Wiley. p. 3. ISBN 9780470390634
- Hájek, Alan (August 28, 2019). "Interpretations of Probability". Stanford Encyclopedia of Philosophy. Retrieved November 17, 2019

Conditional Probability

3

- **Regular Conditional Probability**
- **Conditional Probability Table**
- **Conditional Probability Distribution**
- **Conditional Expectation**
- **Non-commutative Conditional Expectation**
- **Lewis's Triviality Result**
- **Cue Validity**
- **Conditional Variance**

A type of probability in which the probability of an event occurs with a relationship to one or more events is known as conditional probability. It includes conditional probability table, conditional expectation, Lewis's triviality result, conditional variance, etc. This chapter delves into conditional probability for an in-depth understanding of the subject.

Conditional probability is the probability of an event occurring given that the other event has already occurred. The concept is one of the quintessential concepts in probability theory.

Note that the conditional probability does not state that there is always a causal relationship between the two events, as well as it does not indicate that both events occur simultaneously.

Also, the concept of conditional probability is primarily related to the Bayes' theorem, which is one of the most influential theories in statistics.

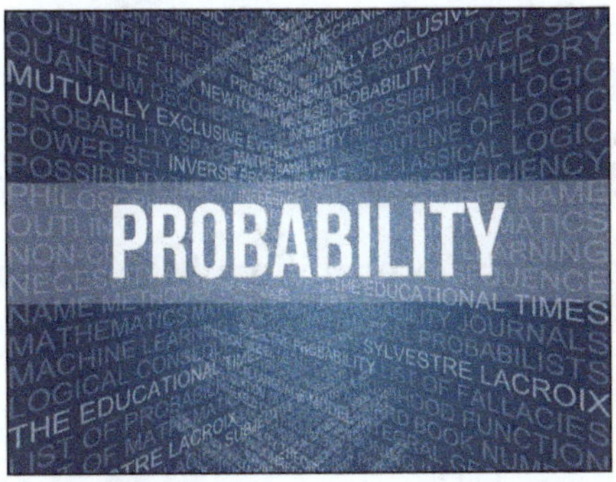

Formula for Conditional Probability,

$$P(A|B) = \frac{P(A \cap B)}{P(B)}$$

Where:

- P(A|B): The conditional probability; the probability of event A occurring given that event B has already occurred.
- P(A ∩ B): The joint probability of events A and B; the probability that both events A and B occur at the same time.
- P(B): The probability of event B.

The formula above is applied to the calculation of the conditional probability of events that are neither independent nor mutually exclusive.

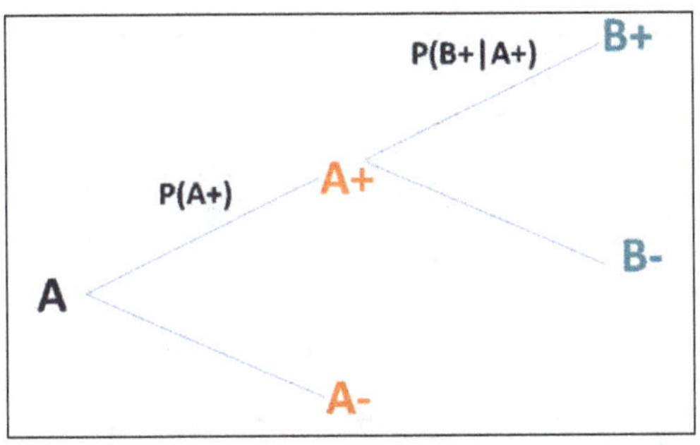

Another way of calculating conditional probability is by using the Bayes' theorem. The theorem can be used to determine the conditional probability of event A, given that the event B has occurred by knowing the conditional probability of event B, given the event A has occurred, as well as the individual probabilities of the event A and B. Mathematically, the Bayes' theorem can be denoted in the following way:

$$P(A|B) = \frac{P(B|A)P(A)}{P(B)}$$

Finally, conditional probabilities can be found using a tree diagram. In the tree diagram, the probabilities in each branch are conditional.

Conditional Probability for Independent Events

Two events are independent if the probability of the outcome of one event does not influence the probability of the outcome of another event. Due to this reason, the conditional probability of two independent events A and B is:

$$P(A|B) = P(A)$$

$$P(B|A) = P(B)$$

Conditional Probability for Mutually Exclusive Events

In probability theory, mutually exclusive events are the events that cannot occur simultaneously. In other words, if one event has already occurred, another can event cannot occur. Thus, the conditional probability of the mutually exclusive events is always zero.

$$P(A|B) = 0$$

$$P(B|A) = 0$$

Conditioning on an Event

Kolmogorov Definition

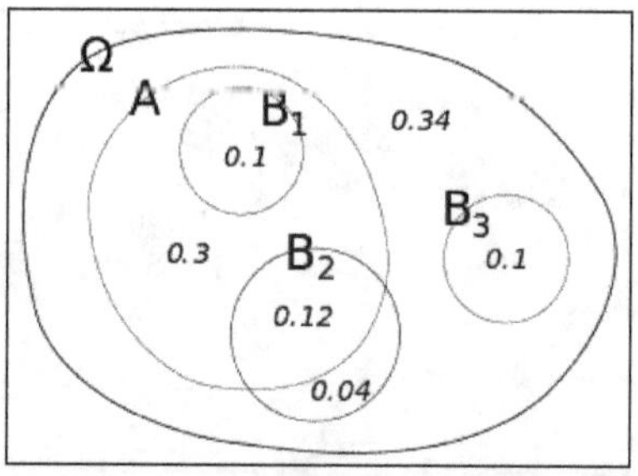

Illustration of conditional probabilities with an Euler diagram. The unconditional probability $P(A) = 0.30 + 0.10 + 0.12 = 0.52$. However, the conditional probability $P(A|B_1) = 1$, $P(A|B_2) = 0.12 \div (0.12 + 0.04) = 0.75$, and $P(A|B_3) = 0$.

Given two events *A* and *B*, from the sigma-field of a probability space, with the unconditional probability of B (that is, of the event *B* occurring) being greater than zero, $P(B) > 0$, the conditional probability of *A* given *B* is defined as the quotient of the probability of the joint of events *A* and *B*, and the probability of *B*:

$$P(A \mid B) = \frac{P(A \cap B)}{P(B)},$$

where $P(A \cap B)$ is the probability that both events *A* and *B* occur. This may be visualized as restricting the sample space to situations in which *B* occurs. The logic behind this equation is that if the possible outcomes for *A* and *B* are restricted to those in which *B* occurs, this set serves as the new sample space.

Note that this is a definition but not a theoretical result. We just denote the quantity $\frac{P(A \cap B)}{P(B)}$ as $P(A \mid B)$ and call it the conditional probability of *A* given *B*.

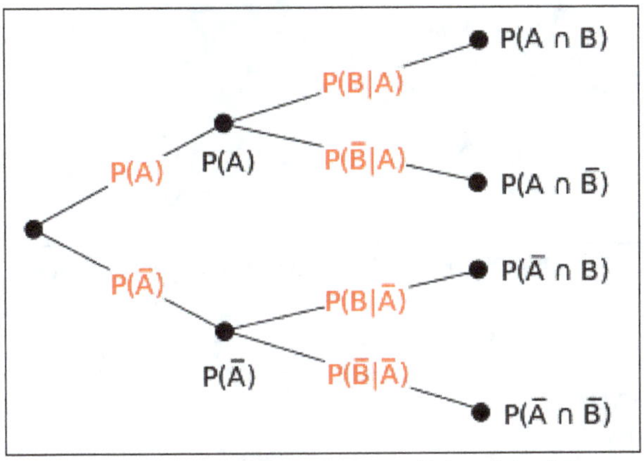

On a tree diagram, branch probabilities are conditional on the event associated with the parent node. (Here the overbars indicate that the event does not occur).

As an Axiom of Probability

Some authors, such as de Finetti, prefer to introduce conditional probability as an axiom of probability:

$$P(A \cap B) = P(A \mid B)P(B)$$

Although mathematically equivalent, this may be preferred philosophically; under major probability interpretations such as the subjective theory, conditional probability is considered a primitive entity. Further, this "multiplication axiom" introduces a symmetry with the summation axiom for mutually exclusive events:

$$P(A \cup B) = P(A) + P(B) - \cancel{P(A \cap B)}^{0}$$

As the Probability of a Conditional Event

Conditional probability can be defined as the probability of a conditional event A_B. Assuming that the experiment underlying the events A and B is repeated, the Goodman–Nguyen–van Fraassen conditional event can be defined as,

$$A_B = \bigcup_{i \ge 1} \left(\bigcap_{j<i} \bar{B}_j, A_i B_i \right).$$

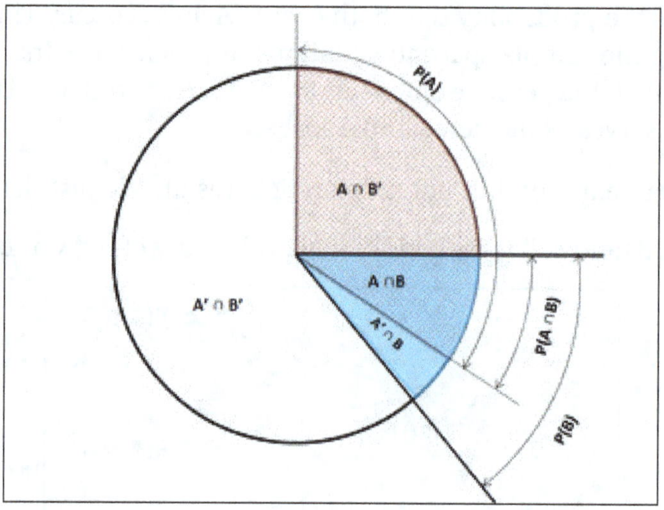

Venn Pie Chart describing conditional probabilities.

It can be shown that,

$$P(A_B) = \frac{P(A \cap B)}{P(B)}$$

which meets the Kolmogorov definition of conditional probability. Note that the equation $P(A_B) = P(A \cap B) / P(B)$ is a theoretical result and not a definition. The definition via conditional events can be understood directly in terms of the Kolmogorov axioms and is particularly close to the Kolmogorov interpretation of probability in terms of experimental data. For example, conditional events can be repeated themselves leading to a generalized notion of conditional event $A_{B(n)}$. It can be shown that the sequence $(A_{B(n)})_{n\geq 1}$ is i.i.d., which yields a strong law of large numbers for conditional probability:

$$P\left(\lim_{n\to\infty} \overline{A}_B^n = P(A \mid B)\right) = 100\%$$

Measure-theoretic Definition

If *P(B)* = 0, then according to the simple definition, *P(A|B)* is undefined. However, it is possible to define a conditional probability with respect to a σ-algebra of such events (such as those arising from a continuous random variable).

For example, if *X* and *Y* are non-degenerate and jointly continuous random variables with density $f_{X,Y}(x, y)$ then, if *B* has positive measure,

$$P(X \in A \mid Y \in B) = \frac{\int_{y \in B} \int_{x \in A} f_{X,Y}(x,y)\,dx\,dy}{\int_{y \in B} \int_{x \in \mathbb{R}} f_{X,Y}(x,y)\,dx\,dy}.$$

The case where B has zero measure is problematic. For the case that $B = \{y_0\}$, representing a single point, the conditional probability could be defined as,

$$P(X \in A \mid Y = y_o) = \frac{\int_{x \in A} f_{X,Y}(x, y_o) dx}{\int_{x \in \mathbb{R}} f_{X,Y}(x, y_o) dx},$$

however this approach leads to the Borel–Kolmogorov paradox. The more general case of zero measure is even more problematic, as can be seen by noting that the limit, as all δy_i approach zero, of,

$$P\left(X \in A \mid Y \in \bigcup_i [y_i, y_i + \delta y_i]\right) \cong \frac{\sum_i \int_{x \in A} f_{X,Y}(x, y_i) dx \, \delta y_i}{\sum_i \int_{x \in \mathbb{R}} f_{X,Y}(x, y_i) dx \, \delta y_i},$$

depends on their relationship as they approach zero.

Conditioning on a Random Variable

Let X be a random variable; we assume for the sake of presentation that X is discrete, that is, X takes on only finitely many values x. Let A be an event. The conditional probability of A given X is defined as the random variable, written P(A|X), that takes on the value,

$$P(A \mid X = x)$$

whenever,

$$X = x.$$

More formally,

$$P(A \mid X)(\omega) = P(A \mid X = X(\omega)).$$

The conditional probability P(A|X) is a function of X: e.g., if the function g is defined as,

$$g(x) = P(A \mid X = x),$$

then,

$$P(A \mid X) = g \circ X.$$

Note that P(A|X) and X are now both random variables. From the law of total probability, the expected value of P(A|X) is equal to the unconditional probability of A.

Partial Conditional Probability

The partial conditional probability $P(A \mid B_1 \equiv b_1, \ldots, B_m \equiv b_m)$ is about the probability of event A given that each of the condition events B_i has occurred to a degree b_i (degree of belief, degree of experience) that might be different from 100%. Frequentistically, partial conditional probability makes sense, if the conditions are tested in experiment repetitions of appropriate length n. Such n-bounded partial conditional probability can be defined as the conditionally expected average occurrence of event A in testbeds of length n that adhere to all of the probability specifications $B_i \equiv b_i$, i.e.:

$$P^n(A \mid B_1 \equiv b_1, \ldots, B_m \equiv b_m) = E(\bar{A}^n \mid \bar{B}_1^n = b_1, \ldots, \bar{B}_m^n = b_m)$$

Based on that, partial conditional probability can be defined as,

$$P(A \mid B_1 \equiv b_1, \ldots, B_m \equiv b_m) = \lim_{n \to \infty} P^n(A \mid B_1 \equiv b_1, \ldots, B_m \equiv b_m),$$

where $b_i n \in \mathbb{N}$,

Jeffrey conditionalization is a special case of partial conditional probability in which the condition events must form a partition:

$$P(A \mid B_1 \equiv b_1, \ldots, B_m \equiv b_m) = \sum_{i=1}^{m} b_i P(A \mid B_i)$$

Suppose that somebody secretly rolls two fair six-sided dice, and we wish to compute the probability that the face-up value of the first one is 2, given the information that their sum is no greater than 5:

- Let D_1 be the value rolled on die 1.
- Let D_2 be the value rolled on die 2.

Probability that $D_1 = 2$

Table below shows the sample space of 36 combinations of rolled values of the two dice, each of which occurs with probability 1/36, with the numbers displayed in the red and dark gray cells being $D_1 + D_2$.

$D_1 = 2$ in exactly 6 of the 36 outcomes; thus $P(D_1 = 2) = \frac{6}{36} = \frac{1}{6}$:

+	D2						
	1	2	3	4	5	6	
D_1	1	2	3	4	5	6	7
	2	3	4	5	6	7	8

	3	4	5	6	7	8	9
	4	5	6	7	8	9	10
	5	6	7	8	9	10	11
	6	7	8	9	10	11	12

Probability that $D_1 + D_2 \leq 5$

Table below shows that $D_1 + D_2 \leq 5$ for exactly 10 of the 36 outcomes, thus $P(D_1 + D_2 \leq 5) = {}^{10}\!/_{36}$:

+	D_2						
	1	2	3	4	5	6	
D_1	1	2	3	4	5	6	7
	2	3	4	5	6	7	8
	3	4	5	6	7	8	9
	4	5	6	7	8	9	10
	5	6	7	8	9	10	11
	6	7	8	9	10	11	12

Probability that $D_1 = 2$ Given that $D_1 + D_2 \leq 5$

Table below shows that for 3 of these 10 outcomes, $D_1 = 2$.

Thus, the conditional probability $P(D_1 = 2 \mid D_1+D_2 \leq 5) = {}^{3}\!/_{10} = 0.3$:

+		D_2					
	1	2	3	4	5	6	
D_1	1	2	3	4	5	6	7
	2	3	4	5	6	7	8
	3	4	5	6	7	8	9
	4	5	6	7	8	9	10
	5	6	7	8	9	10	11
	6	7	8	9	10	11	12

Here, in the earlier notation for the definition of conditional probability, the conditioning event B is that $D_1 + D_2 \leq 5$, and the event *A* is $D_1 = 2$. We have,

$$P(A \mid B) = \frac{P(A \cap B)}{P(B)} = \frac{3/36}{10/36} = \frac{3}{10},$$

as seen in the table.

Use in Inference

In statistical inference, the conditional probability is an update of the probability of an event based on new information. Incorporating the new information can be done as follows:

- Let A, the event of interest, be in the sample space, say (X,P).
- The occurrence of the event A knowing that event B has or will have occurred, means the occurrence of A as it is restricted to B, i.e. $A \cap B$.
- Without the knowledge of the occurrence of B, the information about the occurrence of A would simply be P(A).
- The probability of A knowing that event B has or will have occurred, will be the probability of $A \cap B$ relative to P(B), the probability that B has occurred.
- This results in $P(A \mid B) = P(A \cap B) / P(B)$ whenever P(B) > 0 and 0 otherwise.

This approach results in a probability measure that is consistent with the original probability measure and satisfies all the Kolmogorov axioms. This conditional probability measure also could have resulted by assuming that the relative magnitude of the probability of *A* with respect to *X* will be preserved with respect to *B*.

The wording "evidence" or "information" is generally used in the Bayesian interpretation of probability. The conditioning event is interpreted as evidence for the conditioned event. That is, P(A) is the probability of A before accounting for evidence E, and P(A|E) is the probability of A after having accounted for evidence E or after having updated P(A). This is consistent with the frequentist interpretation, which is the first definition given above.

Common Fallacies

Assuming Conditional Probability is of Similar Size to its Inverse

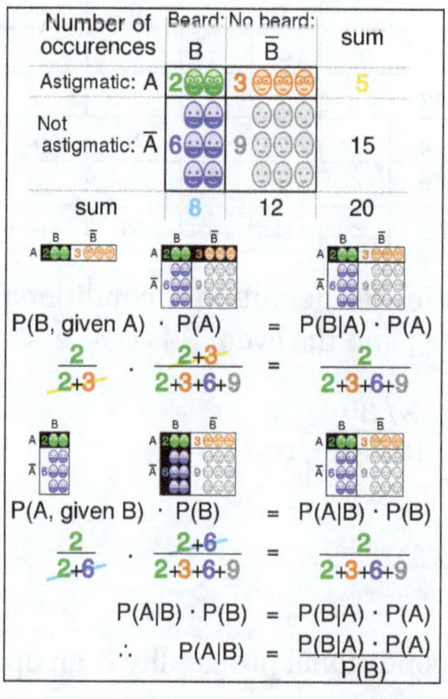

A geometric visualisation of Bayes' theorem: In the table, the values 2, 3, 6 and 9 give the relative weights of each corresponding condition and case. The figures denote the cells of the table involved in each metric, the probability being the fraction of each figure that is shaded. This shows that P(A|B) P(B) = P(B|A) P(A) i.e. $P(A \mid B) = \frac{P(B \mid A)P(A)}{P(B)}$.

Similar reasoning can be used to show that $P(\bar{A} \mid B) = \frac{P(B \mid \bar{A})\, P(\bar{A})}{P(B)}$ etc.

In general, it cannot be assumed that P(A|B) ≈ P(B|A). This can be an insidious error, even for those who are highly conversant with statistics. The relationship between *P(A|B)* and *P(B|A)* is given by Bayes' theorem:

$$P(B \mid A) = \frac{P(A \mid B)P(B)}{P(A)}$$

$$\Leftrightarrow \frac{P(B \mid A)}{P(A \mid B)} = \frac{P(B)}{P(A)}$$

That is, P(A|B) ≈ P(B|A) only if P(B)/P(A) ≈ 1, or equivalently, P(A) ≈ P(B).

Assuming Marginal and Conditional Probabilities are of Similar Size

In general, it cannot be assumed that P(A) ≈ P(A|B). These probabilities are linked through the law of total probability:

$$P(A) = \sum_n P(A \cap B_n) = \sum_n P(A \mid B_n)P(B_n).$$

where the events (B_n) form a countable partition of Ω.

This fallacy may arise through selection bias. For example, in the context of a medical claim, let S_C be the event that a sequela (chronic disease) S occurs as a consequence of circumstance (acute condition) C. Let H be the event that an individual seeks medical help. Suppose that in most cases, C does not cause S so $P(S_C)$ is low. Suppose also that medical attention is only sought if S has occurred due to C. From experience of patients, a doctor may therefore erroneously conclude that $P(S_C)$ is high. The actual probability observed by the doctor is $P(S_C|H)$.

Over- or Under-weighting Priors

Not taking prior probability into account partially or completely is called base rate neglect. The reverse, insufficient adjustment from the prior probability is conservatism.

Formal Derivation

Formally, P(A | B) is defined as the probability of A according to a new probability function on the sample space, such that outcomes not in B have probability 0 and that it is consistent with all original probability measures.

Let Ω be a sample space with elementary events $\{\omega\}$. Suppose we are told the event $B \subseteq \Omega$ has occurred. A new probability distribution (denoted by the conditional notation) is to be assigned on $\{\omega\}$ to reflect this. For events in *B*, it is reasonable to assume that the relative magnitudes of the probabilities will be preserved. For some constant scale factor α, the new distribution will therefore satisfy:

$$\omega \in B : P(\omega \mid B) = \alpha P(\omega)$$
$$\omega \notin B : P(\omega \mid B) = 0$$
$$\sum_{\omega \in \Omega} P(\omega \mid B) = 1.$$

Substituting above equation to select α:

$$\begin{aligned} 1 &= \sum_{\omega \in \Omega} P(\omega \mid B) \\ &= \sum_{\omega \in B} P(\omega \mid B) + \cancelto{0}{\sum_{\omega \notin B} P(\omega \mid B)} \\ &= \alpha \sum_{\omega \in B} P(\omega) \\ &= \alpha \cdot P(B) \\ \Rightarrow \alpha &= \frac{1}{P(B)} \end{aligned}$$

So the new probability distribution is,

$$\omega \in B : P(\omega \mid B) = \frac{P(\omega)}{P(B)}$$
$$\omega \notin B : P(\omega \mid B) = 0$$

Now for a general event *A*,

$$\begin{aligned} P(A \mid B) &= \sum_{\omega \in A \cap B} P(\omega \mid B) + \cancelto{0}{\sum_{\omega \in A \cap B^c} P(\omega \mid B)} \\ &= \sum_{\omega \in A \cap B} \frac{P(\omega)}{P(B)} \\ &= \frac{P(A \cap B)}{P(B)} \end{aligned}$$

Regular Conditional Probability

Regular conditional probability is a concept that has developed to overcome certain difficulties in formally defining conditional probabilities for continuous probability distributions. It is defined as an alternative probability measure conditioned on a particular value of a random variable.

Normally we define the conditional probability of an event A given an event B as:

$$P(A \mid B) = \frac{P(A \cap B)}{P(B)}.$$

The difficulty with this arises when the event B is too small to have a non-zero probability. For example, suppose we have a random variable *X* with a uniform distribution on $[0,1]$, and *B* is the event that $X = 2/3$. Clearly, the probability of *B*, in this case, is $P(B) = 0$, but nonetheless we would still like to assign meaning to a conditional probability such as $P(A \mid X = 2/3)$. To do so rigorously requires the definition of a regular conditional probability.

Let $(\Omega, \mathcal{F}, P)$ be a probability space, and let $T : \Omega \to E$ be a random variable, defined as a Borel-measurable function from Ω to its state space $(E, \mathcal{E})$. Then a regular conditional probability is defined as a function $\nu : E \times \mathcal{F} \to [0,1]$, called a "transition probability", where $\nu(x, A)$ is a valid probability measure (in its second argument) on $\mathcal{F}$ for all $x \in E$ and a measurable function in *E* (in its first argument) for all $A \in \mathcal{F}$, such that for all $A \in \mathcal{F}$ and all $B \in \mathcal{E}$,

$$P\left(A \cap T^{-1}(B)\right) = \int_B \nu(x, A) P\left(T^{-1}(dx)\right).$$

To express this in our more familiar notation:

$$P(A \mid T = x) = \nu(x, A),$$

where $x \in \operatorname{supp} T$, i.e. the topological support of the pushforward measure $T_* P = P\left(T^{-1}(\cdot)\right)$. As can be seen from the integral above, the value of ν for points *x* outside the support of the random variable is meaningless; its significance as a conditional probability is strictly limited to the support of T.

The measurable space $(\Omega, \mathcal{F})$ is said to have the regular conditional probability property if for all probability measures P on $(\Omega, \mathcal{F})$, all random variables on $(\Omega, \mathcal{F}, P)$ admit a regular conditional probability. A Radon space, in particular, has this property.

Alternate Definition

Consider a Radon space Ω (that is a probability measure defined on a Radon space endowed with the Borel sigma-algebra) and a real-valued random variable T. As discussed above, in this case there exists a regular conditional probability with respect to T. Moreover, we can alternatively define the regular conditional probability for an event A given a particular value t of the random variable T in the following manner:

$$P(A \mid T=t) = \lim_{U \supset \{T=t\}} \frac{P(A \cap U)}{P(U)},$$

where the limit is taken over the net of open neighborhoods U of t as they become smaller with respect to set inclusion. This limit is defined if and only if the probability space is Radon, and only in the support of T. This is the restriction of the transition probability to the support of T. To describe this limiting process rigorously:

For every $\epsilon > 0$, there exists an open neighborhood U of the event {T = t}, such that for every open *V* with $\{T=t\} \subset V \subset U$,

$$\left| \frac{P(A \cap V)}{P(V)} - L \right| < \epsilon,$$

where $L = P(A \mid T=t)$ is the limit.

Example: To continue with our motivating example above, we consider a real-valued random variable *X* and write,

$$P(A \mid X=x_0) = v(x_0, A) = \lim_{\epsilon \to 0+} \frac{P(A \cap \{x_0 - \epsilon < X < x_0 + \epsilon\})}{P(\{x_0 - \epsilon < X < x_0 + \epsilon\})},$$

(where $x_0 = 2/3$ for the example given.) This limit, if it exists, is a regular conditional probability for *X*, restricted to $\operatorname{supp} X$.

In any case, it is easy to see that this limit fails to exist for x_0 outside the support of *X*: since the support of a random variable is defined as the set of all points in its state space whose every neighborhood has positive probability, for every point x_0 outside the support of *X* (by definition) there will be an $\epsilon > 0$ such that $P(\{x_0 - \epsilon < X < x_0 + \epsilon\}) = 0$.

Thus if *X* is distributed uniformly on $[0,1]$, it is truly meaningless to condition a probability on "$X = 3/2$".

Conditional Probability Table

In statistics, the conditional probability table (CPT) is defined for a set of discrete and mutually dependent random variables to display conditional probabilities of a single variable with respect to the others (i.e., the probability of each possible value of one variable if we know the values taken on by the other variables). For example, assume there are three random variables x_1, x_2, x_3 where each has K states. Then, the conditional probability table of x_1 provides the conditional probability values $P(x_1 = a_k \mid x_2, x_3)$ – where the vertical bar | means "given the values of" – for each of the *K* possible values a_k of the variable x_1 and for each possible combination of values of x_2, x_3. This table has K^3 cells. In general, for M variables $x_1, x_2, \ldots, x_M$ with K_i states for each variable x_i, the CPT for any one of them has the number of cells equal to the product $K_1 K_2 \cdots K_M$.

A conditional probability table can be put into matrix form. As an example with only two variables, the values of $P(x_1 = a_k \mid x_2 = b_j) = T_{kj}$, with *k* and *j* ranging over *K* values, create a *K*×*K* matrix. This matrix is a stochastic matrix since the columns sum to 1; i.e. $\sum_k T_{kj} = 1$ for all *j*. For example, suppose that two binary variables *x* and *y* have the joint probability distribution given in this table:

	x=0	x=1	P(y)
y=0	4/9	1/9	5/9
y=1	2/9	2/9	4/9
P(x)	6/9	3/9	1

Each of the four central cells shows the probability of a particular combination of x and y values. The first column sum is the probability that x =0 and y equals any of the values it can have – that is, the column sum 6/9 is the marginal probability that x=0. If we want to find the probability that y=0 given that x=0, we compute the fraction of the probabilities in the x=0 column that have the value y=0, which is 4/9 ÷ 6/9 = 4/6. Likewise, in the same column we find that the probability that y=1 given that x=0 is 2/9 ÷ 6/9 = 2/6. In the same way, we can also find the conditional probabilities for y equalling 0 or 1 given that x=1. Combining these pieces of information gives us this table of conditional probabilities for y:

	x=0	x=1
P(y=0 given x)	4/6	1/3
P(y=1 given x)	2/6	2/3
Sum	1	1

With more than one conditioning variable, the table would still have one row for each potential value of the variable whose conditional probabilities are to be given, and there would be one column for each possible combination of values of the conditioning variables.

Moreover, the number of columns in the table could be substantially expanded to display the probabilities of the variable of interest conditional on specific values of only some, rather than all, of the other variables.

Conditional Probability Distribution

In probability theory and statistics, given two jointly distributed random variables X and Y, the conditional probability distribution of Y given X is the probability distribution of Y when X is known to be a particular value; in some cases the conditional probabilities may be expressed as functions containing the unspecified value x of X as a parameter. When both X and Y are categorical variables, a conditional probability table is typically used to represent the conditional probability. The conditional distribution contrasts with the marginal distribution of a random variable, which is its distribution without reference to the value of the other variable.

If the conditional distribution of Y given X is a continuous distribution, then its probability density function is known as the conditional density function. The properties of a conditional distribution, such as the moments, are often referred to by corresponding names such as the conditional mean and conditional variance.

More generally, one can refer to the conditional distribution of a subset of a set of more than two variables; this conditional distribution is contingent on the values of all the remaining variables, and if more than one variable is included in the subset then this conditional distribution is the conditional joint distribution of the included variables.

Conditional Discrete Distributions

For discrete random variables, the conditional probability mass function of Y given $X = x$ can be written according to its definition as:

$$p_{Y|X}(y \mid x) \triangleq P(Y = y \mid X = x) = \frac{P(\{X = x\} \cap \{Y = y\})}{P(X = x)}$$

Due to the occurrence of $P(X = x)$ in a denominator, this is defined only for non-zero (hence strictly positive) $P(X = x)$.

The relation with the probability distribution of X given Y is:

$$P(Y = y \mid X = x)P(X = x) = P(\{X = x\} \cap \{Y = y\}) = P(X = x \mid Y = y)P(Y = y).$$

Consider the roll of a fair die and let $X = 1$ if the number is even (i.e. 2, 4, or 6) and

$X = 0$ otherwise. Furthermore, let $Y = 1$ if the number is prime (i.e. 2, 3, or 5) and $Y = 0$ otherwise.

	1	2	3	4	5	6
X	0	1	0	1	0	1
Y	0	1	1	0	1	0

Then the unconditional probability that $X = 1$ is 3/6 = 1/2 (since there are six possible rolls of the die, of which three are even), whereas the probability that $X = 1$ conditional on $Y = 1$ is 1/3 (since there are three possible prime number rolls—2, 3, and 5—of which one is even).

Conditional Continuous Distributions

Similarly for continuous random variables, the conditional probability density function of Y given the occurrence of the value x of X can be written as,

$$f_{Y|X}(y \mid x) = \frac{f_{X,Y}(x,y)}{f_X(x)}$$

where $f_{X,Y}(x,y)$ gives the joint density of X and Y, while $f_X(x)$ gives the marginal density for X. Also in this case it is necessary that $f_X(x) > 0$.

The relation with the probability distribution of X given Y is given by:

$$f_{Y|X}(y \mid x) f_X(x) = f_{X,Y}(x,y) = f_{X|Y}(x \mid y) f_Y(y).$$

The concept of the conditional distribution of a continuous random variable is not as intuitive as it might seem: Borel's paradox shows that conditional probability density functions need not be invariant under coordinate transformations.

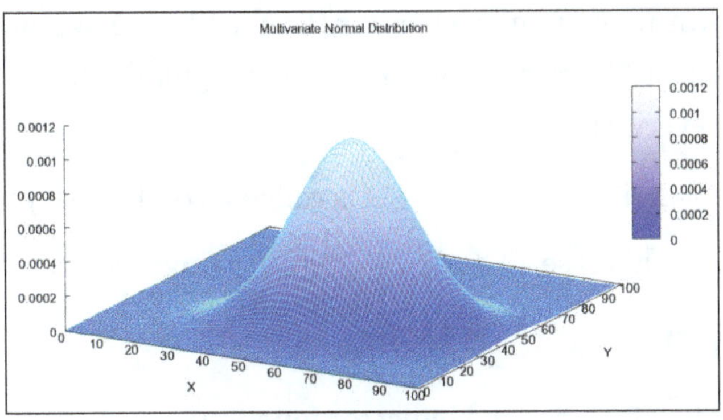

Bivariate normal joint density.

The graph shows a bivariate normal joint density for random variables X and Y. To see the distribution of Y conditional on $X = 70$, one can first visualize the line $X = 70$ in the

X, Y. plane, and then visualize the plane containing that line and perpendicular to the X, Y plane. The intersection of that plane with the joint normal density, once rescaled to give unit area under the intersection, is the relevant conditional density of Y.

$$Y \mid X = 70 \sim \mathcal{N}\left(\mu_1 + \frac{\sigma_1}{\sigma_2}\rho(70 - \mu_2), (1-\rho^2)\sigma_1^2\right).$$

Relation to Independence

Random variables X, Y are independent if and only if the conditional distribution of Y given X is, for all possible realizations of X, equal to the unconditional distribution of Y. For discrete random variables this means $P(Y = y \mid X = x) = P(Y = y)$ for all possible y and x with $P(X = x) > 0$. For continuous random variables X and Y, having a joint density function, it means $f_Y(y \mid X = x) = f_Y(y)$ for all possible y and x with $f_X(x) > 0$.

Properties

Seen as a function of y for given x, $P(Y = y \mid X = x)$ is a probability mass function and so the sum over all y (or integral if it is a conditional probability density) is 1. Seen as a function of x for given y, it is a likelihood function, so that the sum over all x need not be 1.

Measure-theoretic Formulation

Let $(\Omega, \mathcal{F}, P)$ be a probability space, $\mathcal{G} \subseteq \mathcal{F}$ a σ-field in $\mathcal{F}$, and $X : \Omega \to \mathbb{R}$ a real-valued random variable (measurable with respect to the Borel σ-field $\mathcal{R}^1$ on $\mathbb{R}$). Given $A \in \mathcal{F}$,, the Radon-Nikodym theorem implies that there is a $\mathcal{G}$-measurable integrable random variable $P(A \mid \mathcal{G}) : \Omega \to \mathbb{R}$ so that $\int_G P(A \mid \mathcal{G})(\omega) dP(\omega) = P(A \cap G)$ for every $G \in \mathcal{G}$, and such a random variable is uniquely defined up to sets of probability zero. Further, it can then be shown that there exists a function $\mu : \mathcal{R}^1 \times \Omega \to \mathbb{R}$ such that $\mu(\cdot, \omega)$ is a probability measure on $\mathcal{R}^1$ for each $\omega \in \Omega$ (i.e., it is regular) and $\mu(H, \cdot) = P(X^{-1}(H) \mid \mathcal{G})$ (almost surely) for every $H \in \mathcal{R}^1$.

For any $\omega \in \Omega$, the function $\mu(\cdot, \omega) : \mathcal{R}^1 \to \mathbb{R}$ is called a conditional probability distribution of X given $\mathcal{G}$. In this case, $E[X \mid \mathcal{G}] = \int_{-\infty}^{\infty} x \mu(dx, \cdot)$ almost surely.

Relation to Conditional Expectation

For any event $A \in \mathcal{A} \supseteq \mathcal{B}$, define the indicator function:

$$1_A(\omega) = \begin{cases} 1 & \text{if } \omega \in A, \\ 0 & \text{if } \omega \notin A, \end{cases}$$

which is a random variable. Note that the expectation of this random variable is equal to the probability of A itself:

$$E(1_A) = P(A).$$

Then the conditional probability given $\mathcal{B}$ is a function $P(\cdot \mid \mathcal{B}): \mathcal{A} \times \Omega \to (0,1)$ such that $P(A \mid \mathcal{B})$ is the conditional expectation of the indicator function for A:

$$P(A \mid \mathcal{B}) = E(1_A \mid \mathcal{B})$$

In other words, $P(A \mid \mathcal{B})$ is a $\mathcal{B}$-measurable function satisfying

$$\int_B P(A \mid \mathcal{B})(\omega) \, dP(\omega) = P(A \cap B) \qquad \text{for all} \quad A \in \mathcal{A}, B \in \mathcal{B}.$$

A conditional probability is regular if $P(\cdot \mid \mathcal{B})(\omega)$ is also a probability measure for all $\omega \in \Omega$. An expectation of a random variable with respect to a regular conditional probability is equal to its conditional expectation.

- For the trivial sigma algebra $\mathcal{B} = \{\varnothing, \Omega\}$ the conditional probability is a constant function, $P(A \mid \{\varnothing, \Omega\}) \equiv P(A)$.
- For $A \in \mathcal{B}$, as outlined above, $P(A \mid \mathcal{B}) = 1_A$.

Conditional Expectation

In probability theory, the conditional expectation, conditional expected value, or conditional mean of a random variable is its expected value – the value it would take "on average" over an arbitrarily large number of occurrences – given that a certain set of "conditions" is known to occur. If the random variable can take on only a finite number of values, the "conditions" are that the variable can only take on a subset of those values. More formally, in the case when the random variable is defined over a discrete probability space, the "conditions" are a partition of this probability space.

With multiple random variables, for one random variable to be mean independent of all others both individually and collectively means that each conditional expectation equals the random variable's (unconditional) expected value. This always holds if the variables are independent, but mean independence is a weaker condition.

Depending on the nature of the conditioning, the conditional expectation can be either a random variable itself or a fixed value. With two random variables, if the expectation of a random variable X is expressed conditional on another random variable Y without a particular value of Y being specified, then the expectation of X conditional on Y, denoted $E(X \mid Y)$, is a function of the random variable Y and hence is itself a random variable. Alternatively, if the expectation of X is expressed conditional on the occurrence of a particular value of Y, denoted y, then the conditional expectation $E(X \mid Y = y)$ is a fixed value.

This concept generalizes to any probability space using measure theory. In modern probability theory the concept of conditional probability is defined in terms of conditional expectation.

Example: Consider the roll of a fair die and let A = 1 if the number is even (i.e. 2, 4, or 6) and A = 0 otherwise. Furthermore, let B = 1 if the number is prime (i.e. 2, 3, or 5) and B = 0 otherwise.

	1	2	3	4	5	6
A	0	1	0	1	0	1
B	0	1	1	0	1	0

The unconditional expectation of A is $E[A] = (0+1+0+1+0+1)/6 = 1/2$. But the expectation of A conditional on B = 1 (i.e., conditional on the die roll being 2, 3, or 5) is $E[A \mid B = 1] = (1+0+0)/3 = 1/3$, and the expectation of A conditional on B = 0 (i.e., conditional on the die roll being 1, 4, or 6) is $E[A \mid B = 0] = (0+1+1)/3 = 2/3$. Likewise, the expectation of B conditional on A = 1 is $E[B \mid A = 1] = (1+0+0)/3 = 1/3$, and the expectation of B conditional on A = 0 is $E[B \mid A = 0] = (0+1+1)/3 = 2/3$.

Example: Suppose we have daily rainfall data (mm of rain each day) collected by a weather station on every day of the ten–year (3652–day). The unconditional expectation of rainfall for an unspecified day is the average of the rainfall amounts of those 3652 days. The conditional expectation of rainfall for an otherwise unspecified day known to be (conditional on being) in the month of March is the average of daily rainfall over all 310 days of the ten–year period that falls in March. And the conditional expectation of rainfall conditional on days dated March 2 is the average of the rainfall amounts that occurred on the ten days with that specific date.

Conditional Expectation with Respect to an Event

In classical probability theory the conditional expectation of X given an event H (which may be the event Y y for a random variable Y) is the average of X overall outcomes in H, that is,

$$E(X \mid H) = \frac{\sum_{\omega \in H} X(\omega)}{|H|},$$

where $|H|$ is the cardinality of H.

The sum above can be grouped by different values of $X(\omega)$, to get a sum over the range $\mathcal{X}$ of X

$$E(X \mid H) = \sum_{x \in \mathcal{X}} x \frac{|\{\omega \in H \mid X(\omega) = x\}|}{|H|}.$$

In modern probability theory, when H is an event with strictly positive probability, it is possible to give a similar formula. This is notably the case for a discrete random variable Y and for y in the range of Y if the event H is $Y = y$. Let $(\Omega, \mathcal{F}, P)$ be a probability space, X is a random variable on that probability space, and $H \in \mathcal{F}$ an event with strictly positive probability $P(H) > 0$. Then the conditional expectation of X given the event H is,

$$E(X \mid H) = \frac{E(1_H X)}{P(H)} = \int_{\mathcal{X}} x dP(x \mid H),$$

where $\mathcal{X}$ is the range of X and $P(\cdot \mid H)$ is the probability measure defined, for each set A, as $P(A \mid H) = P(A \cap H)/P(H)$, the conditional probability of A given H.

When $P(H) = 0$ (for instance if Y is a continuous random variable and H is the event $Y = y$, this is in general the case), the Borel–Kolmogorov paradox demonstrates the ambiguity of attempting to define the conditional probability knowing the event H. The above formula shows that this problem transposes to the conditional expectation. So instead one only defines the conditional expectation with respect to a σ-algebra or a random variable.

Conditional Expectation with Respect to a Random Variable

If *Y* is a discrete random variable on the same probability space $(\Omega, \mathcal{F}, P)$ having range $\mathcal{Y}$, then the conditional expectation of X with respect to Y is the function $E(X \mid Y)(\cdot)$ of the variable $y \in \mathcal{Y}$ defined by,

$$E(X \mid Y)(y) = E(X \mid Y = y).$$

There is a closely related function from Ω to $\mathcal{Y}$ defined by,

$$E(X \mid \sigma(Y))(\omega) = E(X \mid Y = Y(\omega)).$$

This function, which is different from the previous one, is the conditional expectation of *X* with respect to the σ-algebra generated by *Y*. The two are related by,

$$E(X \mid \sigma(Y)) = E(X \mid Y) \circ Y.$$

where $\circ$ stands for Function composition.

As mentioned above, if Y is a continuous random variable, it is not possible to define $E(X \mid Y)$ by this method. As explained in the Borel–Kolmogorov paradox, we have to specify what limiting procedure produces the set $Y = y$. If the event space Ω has a distance function, then one procedure for doing so is as follows. Define the set $H_y^\varepsilon = \{\omega \mid \| Y(\omega) - y \| < \varepsilon\}$. Assume that each H_y^ε is P-measurable and that $P(H_y^\varepsilon) > 0$ for all $\varepsilon > 0$. Then conditional expectation with respect to H_y^ε is well-defined. Take the limit as ε tends to 0 and define,

$$g(y) = \lim_{\varepsilon \to 0} E(X \mid H_y^\varepsilon).$$

Replacing this limiting process by the Radon–Nikodym derivative yields an analogous definition that works more generally.

Formal Definition

Conditional Expectation with Respect to a Sub-σ-algebra

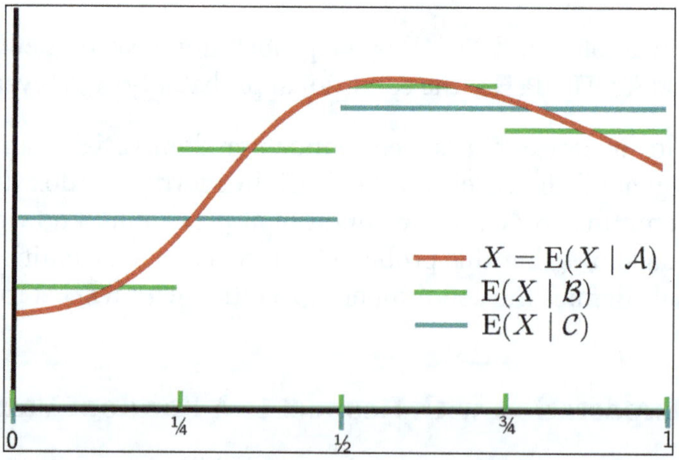

Conditional expectation with respect to a σ-algebra: in this example the probability space $(\Omega, \mathcal{F}, P)$ is the [0, 1] interval with the Lebesgue measure. We define the following σ-algebras: $\mathcal{A} = \mathcal{F}$; $\mathcal{B}$ is the σ-algebra generated by the intervals with end-points 0, ¼, ½, ¾, 1; and $\mathcal{C}$ is the σ-algebra generated by the intervals with end-points 0, ½, 1. Here the conditional expectation is effectively the average over the minimal sets of the σ-algebra.

Consider the following:

- $(\Omega, \mathcal{F}, P)$ is a probability space.
- $X : \Omega \to \mathbb{R}^n$ is a random variable on that probability space with finite expectation.
- $\mathcal{H} \subseteq \mathcal{F}$ is a sub-σ-algebra of $\mathcal{F}$.

Since $\mathcal{H}$ is a sub σ-algebra of $\mathcal{F}$, the function $X:\Omega \to \mathbb{R}^n$ is usually not $\mathcal{H}$-measurable, thus the existence of the integrals of the form $\int_H X dP|_{\mathcal{H}}$, where $H \in \mathcal{H}$ and $P|_{\mathcal{H}}$ is the restriction of P to $\mathcal{H}$, cannot be stated in general. However, the local averages $\int_H X dP$ can be recovered in $(\Omega, \mathcal{H}, P|_{\mathcal{H}})$ with the help of the conditional expectation. A conditional expectation of X given $\mathcal{H}$, denoted as $E(X|\mathcal{H})$, is any $\mathcal{H}$-measurable function $\Omega \to \mathbb{R}^n$ which satisfies:

$$\int_H E(X|\mathcal{H}) dP = \int_H X dP$$

for each $H \in \mathcal{H}$.

The existence of $E(X|\mathcal{H})$ can be established by noting that $\mu^X : F \mapsto \int_F X dP$ for $F \in \mathcal{F}$ is a finite measure on $(\Omega, \mathcal{F})$ that is absolutely continuous with respect to P. If h is the natural injection from $\mathcal{H}$ to $\mathcal{F}$, then $\mu^X \circ h = \mu^X|_{\mathcal{H}}$ is the restriction of μ^X to $\mathcal{H}$ and $P \circ h = P|_{\mathcal{H}}$ is the restriction of P to $\mathcal{H}$. Furthermore, $\mu^X \circ h$ is absolutely continuous with respect to $P \circ h$, because the condition,

$$P \circ h(H) = 0 \Leftrightarrow P(h(H)) = 0$$

implies,

$$\mu^X(h(H)) = 0 \Leftrightarrow \mu^X \circ h(H) = 0.$$

Thus, we have,

$$E(X|\mathcal{H}) = \frac{d\mu^X|_{\mathcal{H}}}{dP|_{\mathcal{H}}} = \frac{d(\mu^X \circ h)}{d(P \circ h)},$$

where the derivatives are Radon–Nikodym derivatives of measures.

Conditional Expectation with Respect to a Random Variable

Consider, in addition to the above,

- A measurable space (U, Σ),
- A random variable $Y : \Omega \to U$.

Let $g : U \to \mathbb{R}^n$ be a Σ-measurable function such that, for every Σ-measurable function $f : U \to \mathbb{R}^n$,

$$\int g(Y) f(Y) dP = \int X f(Y) dP.$$

Then the random variable $g(Y)$, denoted as $E(X \mid Y)$, is a conditional expectation of X given Y.

This definition is equivalent to defining the conditional expectation with respect to the sub-σ-field of $\mathcal{F}$ defined by the pre-image of Σ by Y. If we define,

$$\mathcal{H} = Y^{-1}(\Sigma) = \{Y^{-1}(B) : B \in \Sigma\},$$

then,

$$E(X \mid Y) \circ Y = E(X \mid \mathcal{H}) = \frac{d(\mu^X \circ Y^{-1})}{d(P \circ Y^{-1})} \circ Y$$

- This is not a constructive definition; we are merely given the required property that a conditional expectation must satisfy:
    - The definition of $E(X \mid \mathcal{H})$ may resemble that of $E(X \mid H)$ for an event H but these are very different objects. The former is a $\mathcal{H}$-measurable function $\Omega \to \mathbb{R}^n$, while the latter is an element of $\mathbb{R}^n$. Evaluating the former at any $\omega \in H$ yields the latter.
    - Existence of a conditional expectation function may be proven by the Radon–Nikodym theorem. A sufficient condition is that the (unconditional) expected value for X exists.
    - Uniqueness can be shown to be almost sure: that is, versions of the same conditional expectation will only differ on a set of probability zero.
- The σ-algebra $\mathcal{H}$ controls the "granularity" of the conditioning. A conditional expectation $E(X \mid \quad)$ over a finer (larger) σ-algebra $\mathcal{H}$ retains information about the probabilities of a larger class of events. A conditional expectation over a coarser (smaller) σ-algebra averages over more events.

Conditioning as Factorization

In the definition of conditional expectation that we provided above, the fact that Y is a *real* random element is irrelevant. Let (U, Σ) be a measurable space, where Σ is a σ-algebra on U. A U-valued random element is a measurable function $Y : \Omega \to U$, i.e. $Y^{-1}(B) \in \mathcal{F}$ for all $B \in \Sigma$. The distribution of Y is the probability measure $P_Y : \Sigma \to \mathbb{R}$ defined as the pushforward measure $Y_* P$, that is, such that $P_Y(B) = P(Y^{-1}(B))$.

Theorem: If $X : \Omega \to \mathbb{R}$ is an integrable random variable, then there exists a unique integrable random element $E(X \mid Y) : U \to \mathbb{R}$, defined P_Y almost surely, such that,

$$\int_{Y^{-1}(B)} X \, dP = \int_B E(X \mid Y) \, dP_Y,$$

for all $B \in \Sigma$.

Proof: Let $\mu:\Sigma \to \mathbb{R}$ be such that $\mu(B)=\int_{Y^{-1}(B)} X dP$. Then μ is a signed measure which is absolutely continuous with respect to P_Y. Indeed $P_Y(B)=0$ means exactly that $P(Y^{-1}(B))=0$, and since the integral of an integrable function on a set of probability 0 is 0, this proves absolute continuity. The Radon–Nikodym theorem then proves the existence of a density of μ with respect to P_Y. This density is $E(X \mid Y)$.

Comparing with conditional expectation with respect to sub-σ-algebras, it holds that,

$$E(X \mid Y)\circ Y = E\left(X \mid Y^{-1}(\Sigma)\right).$$

We can further interpret this equality by considering the abstract change of variables formula to transport the integral on the right hand side to an integral over Ω:

$$\int_{Y^{-1}(B)} X dP = \int_{Y^{-1}(B)} \left(E\left(X \mid Y\right)\circ Y\right) dP.$$

The equation means that the integrals of X and the composition $E(X \mid Y)\circ Y$ over sets of the form $Y^{-1}(B)$, for $B \in \Sigma$, are identical.

This equation can be interpreted to say that the following diagram is commutative on average.

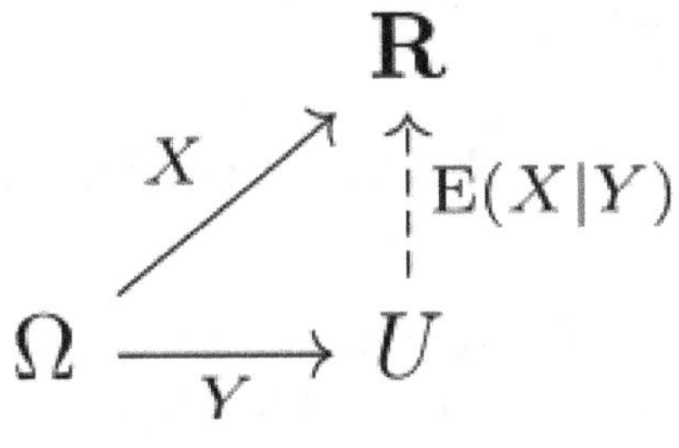

Computation

When X and Y are both discrete random variables, then the conditional expectation of X given the event Y = y can be considered as function of y for y in the range of Y:

$$E(X \mid Y=y)=\sum_{x\in\mathcal{X}} xP(X=x \mid Y=y)=\sum_{x\in\mathcal{X}} x\frac{P(X=x, Y=y)}{P(Y=y)},$$

where $\mathcal{X}$ is the range of X.

If X is a continuous random variable, while Y remains a discrete variable, the conditional expectation is,

$$E(X \mid Y=y)=\int_{\mathcal{X}} x f_X(x \mid Y=y) dx,$$

with $f_X(x \mid Y=y) = \frac{f_{X,Y}(x,y)}{P(Y=y)}$ (where $f_{X,Y}(x, y)$ gives the joint density of X and Y) being the conditional density of X given Y = y.

If both *X* and *Y* are continuous random variables, then the conditional expectation is

$$E(X \mid Y=y) = \int_{\mathcal{X}} x f_{X|Y}(x \mid y) dx,$$

where $f_{X|Y}(x \mid y) = \frac{f_{X,Y}(x,y)}{f_Y(y)}$ (where $f_Y(y)$ gives the density of Y).

Basic Properties

All the following formulas are to be understood in an almost sure sense. The σ-algebra $\mathcal{H}$ could be replaced by a random variable Z.

- Pulling out independent factors:
    - If X is independent of $\mathcal{H}$, then $E(X \mid \mathcal{H}) = E(X)$.

Proof:

Let $B \in \mathcal{H}$. Then X is independent of 1_B, so we get that:

$$\int_B X dP = E(X1_B) = E(X)E(1_B) = E(X)P(B) = \int_B E(X) dP.$$

Thus the definition of conditional expectation is satisfied by the constant random variable E(X), as desired.

- If X is independent of $\sigma(Y,\mathcal{H})$, then $E(XY \mid \mathcal{H}) = E(X)E(Y \mid \mathcal{H})$.. Note that this is not necessarily the case if X is only independent of $\mathcal{H}$ and of Y.
- If X, Y are independent, $\mathcal{G},\mathcal{H}$ are independent, X is independent of $\mathcal{H}$ and Y is independent of $\mathcal{G}$ then $E(E(XY \mid \mathcal{G}) \mid \mathcal{H}) = E(X)E(Y) = E(E(XY \mid \mathcal{H}) \mid \mathcal{G})$.

- Stability:
    - If X is $\mathcal{H}$-measurable, then $E(X \mid \mathcal{H}) = X$.
    - If Z is a random variable, then $E(f(Z) \mid Z) = f(Z)$. In its simplest form, this says $E(Z \mid Z) = Z$.
- Pulling out known factors:
    - If X is $\mathcal{H}$-measurable, then $E(XY \mid \mathcal{H}) = XE(Y \mid \mathcal{H})$.
    - If Z is a random variable, then $E(f(Z)Y \mid Z) = f(Z)E(Y \mid Z)$.
- Law of total expectation: $E(E(X \mid \mathcal{H})) = E(X)$.

- Tower property:
  - For sub-σ-algebras $\mathcal{H}_1 \subset \mathcal{H}_2 \subset \mathcal{F}$ we have $E(E(X \mid \mathcal{H}_2) \mid \mathcal{H}_1) = E(X \mid \mathcal{H}_1)$.
    - A special case is when Z is a $\mathcal{H}$-measurable random variable. Then $\sigma(Z) \subset \mathcal{H}$ and thus $E(E(X \mid \mathcal{H}) \mid Z) = E(X \mid Z)$.
    - Doob martingale property: The above with $Z = E(X \mid \mathcal{H})$ (which is $\mathcal{H}$-measurable), and using also $E(Z \mid Z) = Z$, gives $E(X \mid E(X \mid \mathcal{H})) = E(X \mid \mathcal{H})$.
  - For random variables X, Y we have $E(E(X \mid Y) \mid f(Y)) = E(X \mid f(Y))$.
  - For random variables X, Y, Z we have $E(E(X \mid Y, Z) \mid Y) = E(X \mid Y)$.
- Linearity: We have $E(X_1 + X_2 \mid \mathcal{H}) = E(X_1 \mid \mathcal{H}) + E(X_2 \mid \mathcal{H})$ and $E(aX \mid \mathcal{H}) = aE(X \mid \mathcal{H})$ for $a \in \mathbb{R}$.
- Positivity: If $X \geq 0$ then $E(X \mid \mathcal{H}) \geq 0$.
- Monotonicity: If $X_1 \leq X_2$ then $E(X_1 \mid \mathcal{H}) \leq E(X_2 \mid \mathcal{H})$.
- Monotone convergence: If $0 \leq X_n \uparrow X$ then $E(X_n \mid \mathcal{H}) \uparrow E(X \mid \mathcal{H})$.
- Dominated convergence: If $X_n \to X$ and $|X_n| \leq Y$ with $Y \in L^1$, then $E(X_n \mid \mathcal{H}) \to E(X \mid \mathcal{H})$.
- Fatou's lemma: If $E(\inf_n X_n \mid \mathcal{H}) > -\infty$ then $E(\liminf_{n\to\infty} X_n \mid \mathcal{H}) \leq \liminf_{n\to\infty} E(X_n \mid \mathcal{H})$.
- Jensen's inequality: If $f : \mathbb{R} \to \mathbb{R}$ is a convex function, then $f(E(X \mid \mathcal{H})) \leq E(f(X) \mid \mathcal{H})$.
- Conditional variance: Using the conditional expectation we can define, by analogy with the definition of the variance as the mean square deviation from the average, the conditional variance:
  - Definition: $\mathrm{Var}(X \mid \mathcal{H}) = E\left((X - E(X \mid \mathcal{H}))^2 \mid \mathcal{H}\right)$
  - Algebraic formula for the variance: $\mathrm{Var}(X \mid \mathcal{H}) = E(X^2 \mid \mathcal{H}) - \left(E(X \mid \mathcal{H})\right)^2$
  - Law of total variance: $\mathrm{Var}(X) = E(\mathrm{Var}(X \mid \mathcal{H})) + \mathrm{Var}(E(X \mid \mathcal{H}))$.
- Martingale convergence: For a random variable X, that has finite expectation, we have $E(X \mid \mathcal{H}_n) \to E(X \mid \mathcal{H})$,, if either $\mathcal{H}_1 \subset \mathcal{H}_2 \subset \cdots$ is an increasing series of sub-σ-algebras and $\mathcal{H} = \sigma\left(\bigcup_{n=1}^{\infty} \mathcal{H}_n\right)$ or if $\mathcal{H}_1 \supset \mathcal{H}_2 \supset \cdots$ is a decreasing series of sub-σ-algebras and $\mathcal{H} = \bigcap_{n=1}^{\infty} \mathcal{H}_n$.

- Conditional expectation as L^2-projection: If X, Y are in the Hilbert space of square-integrable real random variables (real random variables with finite second moment) then:
    - For $\mathcal{H}$-measurable Y, we have $E(Y(X - E(X \mid \mathcal{H}))) = 0$, i.e. the conditional expectation $E(X \mid \mathcal{H})$ is in the sense of the $L^2(P)$ scalar product the orthogonal projection from X to the linear subspace of $\mathcal{H}$-measurable functions. (This allows to define and prove the existence of the conditional expectation based on the Hilbert projection theorem).
    - The mapping $X \mapsto E(X \mid \mathcal{H})$ is self-adjoint: $E(X\,E(Y \mid \mathcal{H})) = E\big(E(X \mid \mathcal{H})\,E(Y \mid \mathcal{H})\big) = E(E(X \mid \mathcal{H})Y)$.
- Conditioning is a contractive projection of L^p spaces $L^p(\Omega, \mathcal{F}, P) \rightarrow L^p(\Omega, \mathcal{H}, P)$. I.e., $E\left(|E(X \mid \mathcal{H})|^p\right) \le E\left(|X|^p\right)$ for any $p \ge 1$.
- Doob's conditional independence property: If X, Y are conditionally independent given Z, then $P(X \in B \mid Y, Z) = P(X \in B \mid Z)$ (equivalently, $E(1_{\{X \in B\}} \mid Y, Z) = E(1_{\{X \in B\}} \mid Z)$).

Non-commutative Conditional Expectation

In mathematics, non-commutative conditional expectation is a generalization of the notion of conditional expectation in classical probability. The space of measurable functions on a σ-finite measure space (X, μ) is the canonical example of a commutative von Neumann algebra. For this reason, the theory of von Neumann algebras is sometimes referred to as noncommutative measure theory. The intimate connections of probability theory with measure theory suggest that one may be able to extend the classical ideas in probability to a noncommutative setting by studying those ideas on general von Neumann algebras.

For von Neumann algebras with a faithful normal tracial state, for example finite von Neumann algebras, the notion of conditional expectation is especially useful.

A positive, linear mapping $\mathcal{R}$ of a von Neumann algebra $\mathcal{S}$ onto a von Neumann algebra $\mathcal{R}$ ($\mathcal{S}$ and $\mathcal{R}$ may be general C*-algebras as well) is said to be a *conditional expectation* (of $\mathcal{S}$ onto $\mathcal{R}$) when $\Phi(I) = I$ and $\Phi(R_1 S R_2) = R_1 \Phi(S) R_2$ if $R_1, R_2 \in \mathcal{R}$ and $S \in \mathcal{S}$.

Applications

Sakai's Theorem

Let $\mathcal{B}$ be a C*-subalgebra of the C*-algebra $\mathfrak{A}, \varphi_0$ an idempotent linear mapping of $\mathfrak{A}$ onto $\mathcal{B}$ such that $\| \varphi_0 \| = 1, \mathfrak{A}$ acting on $\mathcal{H}$ the universal representation of $\mathfrak{A}$. Then φ_0 extends uniquely to an ultraweakly continuous idempotent linear mapping φ of $\mathfrak{A}^-$, the weak-operator closure of $\mathfrak{A}$, onto $\mathcal{B}^-$, the weak-operator closure of $\mathcal{B}$.

In the above setting, a result first proved by Tomiyama may be formulated in the following manner.

Theorem: Let $\mathfrak{A}, \mathcal{B}, \varphi, \varphi_0$ be as described above. Then φ is a conditional expectation from $\mathfrak{A}^-$, onto $\mathcal{B}^-$ and φ_0 is a conditional expectation from $\mathfrak{A}$ onto $\mathcal{B}$.

With the aid of Tomiyama's theorem an elegant proof of Sakai's result on the characterization of those C*-algebras that are *-isomorphic to von Neumann algebras may be given.

Lewis's Triviality Result

In the mathematical theory of probability, David Lewis's triviality result is a theorem about the impossibility of systematically equating the conditional probability $P(B \mid A)$ with the probability of a so-called conditional event, $A \rightarrow B$.

Conditional Probability and Conditional Events

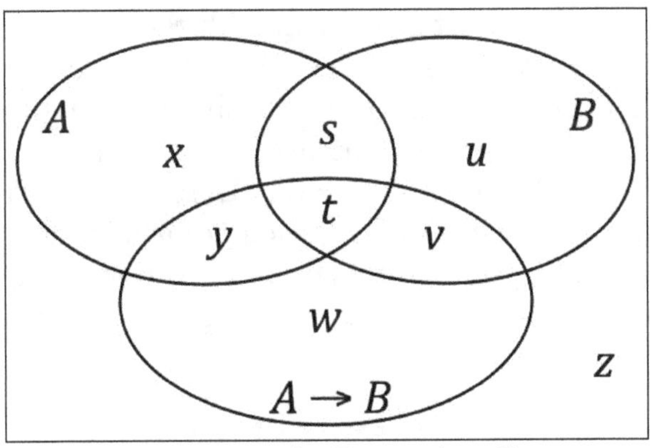

A diagram of A, B, and $A \rightarrow B$.

The statement "The probability that if A, then B, is 20%" means (put intuitively) that event B may be expected to occur in 20% of the outcomes where event A occurs. The standard formal expression of this is $P(B \mid A) = 0.20$, where the conditional probability $P(B \mid A)$ equals, by definition, $P(A \cap B) / P(A)$.

Beginning in the 1960s, several philosophical logicians—most notably Ernest Adams and Robert Stalnaker—floated the idea that one might also write $P(A \to B) = 0.20$ where $A \to B$ is the conditional event "If A, then B". That is, given events A and B, one might suppose there is an event, $A \to B$, such that $P(A \to B)$ could be counted on to equal $P(B \mid A)$, so long as $P(A) > 0$.

Part of the appeal of this move would be the possibility of embedding conditional expressions within more complex constructions. One could write, say, $P(A \cup (B \to C)) = 0.75$, to express someone's high subjective degree of confidence ("75% sure") that either A, or else if B, then C. Compound expressions containing conditional expressions might also be useful in the programming of automated decision-making systems.

How might such a convention be combined with standard probability theory? The most direct extension of the standard theory would be to treat $A \to B$ as an event like any other, i.e., as a set of outcomes. Adding $A \to B$ to the familiar Venn- or Euler diagram of A and B would then result in something like figure above, where $s, t, \ldots, z$ are probabilities allocated to the eight respective regions, such that $s + t + \cdots + z = 1$.

For $P(A \to B)$ to equal $P(B \mid A)$ requires that $t + v + w + y = (s + t)/(s + t + x + y)$, i.e., that the probability inside the $A \to B$ region equal the $A \cap B$ region's proportional share of the probability inside the A region. In general the equality will of course not be true, so that making it reliably true requires a new constraint on probability functions: in addition to satisfying Kolmogorov's probability axioms, they must also satisfy a new constraint, namely that $P(A \to B) = P(B \mid A)$ for any events A and B such that $P(A) > 0$.

Lewis's Result

Lewis pointed out a seemingly fatal problem with the above proposal: assuming a nontrivial set of events, the new, restricted class of P-functions will not be closed under conditioning, the operation that turns probability function P into new function $P_C(\cdot) = P(\cdot \mid C)$, predicated on event C's occurrence. That is, if $P(A \to B) = P(B \mid A)$, it will not in general be true that $P_C(A \to B) = P_C(B \mid A)$ as long as $P(C) > 0$. This implies that if rationality requires having a well-behaved probability function, then a fully rational person (or computing system) would become irrational simply in virtue of learning that arbitrary event C had occurred. Bas van Fraassen called this result "a veritable bombshell".

Lewis's proof is as follows. Let a set of events be non-trivial if it contains two possible events, A and B, that are mutually exclusive but do not together exhaust all possibilities, so that $P(A) > 0$, $P(B) > 0$, $P(A \cap B) = 0$, and $P(A \cup B) < 1$. The existence of two such events implies the existence of the event $A \cup B$, as well, and, if conditional events are admitted, the event $(A \cup B) \to A$. The proof derives a contradiction from the assumption that such a minimally non-trivial set of events exists.

Step 1: Consider the probability of $(A \cup B) \to A$ after conditioning, first on A and then instead on A′:

- Conditioning on A gives $P_A((A \cup B) \to A) = P(((A \cup B) \to A) \cap A) / P(A)$. But also, by the new constraint on P-functions, $P_A((A \cup B) \to A) = P_A((A \cup B) \cap A) / P_A(A \cup B) = P((A \cup B) \cap A \mid A) / P(A \cup B \mid A) = 1/1 = 1$. Therefore, $P(((A \cup B) \to A) \cap A) = P(A)$.
- Conditioning on A′ gives $P_{A'}((A \cup B) \to A) = P(((A \cup B) \to A) \cap A') / P(A')$, But also, $P_{A'}((A \cup B) \to A) = P_{A'}((A \cup B) \cap A) / P_{A'}(A \cup B) = P((A \cup B) \cap A \mid A') / P(A \cup B \mid A') = 0 / P(A \cup B \mid A') = 0$. (The mutual exclusivity of B and A ensures that $P(A \cup B \mid A') \neq 0$). Therefore, $P(((A \cup B) \to A) \cap A') = 0$.

Step: Instantiate the identity $P(X \cap Y) + P(X \cap Y') = P(X)$ as,

$$P(((A \cup B) \to A) \cap A) + P(((A \cap B) \to A) \cap A') = P((A \cup B) \to A).$$

By the results from Step 1, the left side reduces to $P(A)$, while the right side, by the new constraint on P-functions, equals $P((A \cup B) \cap A) / P(A \cup B) = P(A) / P(A \cup B)$. Therefore, $P(A) = P(A) / P(A \cup B)$ which means that $P(A \cup B) = 1$, which contradicts the stipulation that $P(A \cup B) = 1$. This completes the proof.

Graphical Version

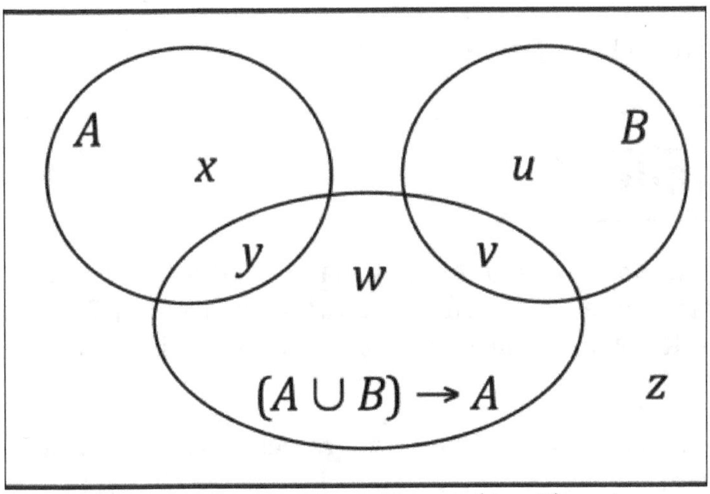

A diagram of disjoint A and B,, and $(A \cup B) \to A$.

A graphical version of the proof starts with figure above, where the A and B from figure are now disjoint and $A \to B$ has been replaced by $(A \cup B) \to A$. By the assumption that A and B are possible, $x + y > 0$ and $u + v > 0$. By the assumption that together A and B do not together exhaust all possibilities, $u + v + x + y < 1$.

And by the new constraint on probability functions, $P((A \cup B) \rightarrow A) = P(A \mid A \cup B) = P(A \cap (A \cup B)) / P(A \cup B) = P(A) / P(A \cup B)$ which means that,

$$y + v + w = \frac{x + y}{x + y + u + v},$$

Conditioning on an event involves zeroing out the probabilities outside the event's region and increasing the probabilities inside the region by a common scale factor. Here, conditioning on A will zero out u, v and w and scale up x and y, to x_A and y_A, respectively, and so,

$$y_A + 0 + 0 = \frac{x_A + y_A}{x_A + y_A + 0 + 0}, \text{which simplifies to } y_A = 1.$$

Conditioning instead on A' will zero out x and y and scale up u, v and w, and so,

$$0 + v_{A'} + w_{A'} = \frac{0 + 0}{0 + 0 + u_A + v_A}, \text{which simplifies to } v_{A'} + w_{A'} = 0.$$

From equation $y_A + 0 + 0 = \dfrac{x_A + y_A}{x_A + y_A + 0 + 0}$,

it follows that $x_A = 0$ and since x_A is the scaled-up value of x, it must also be that $x = 0$ Similarly, from equation $0 + v_{A'} + w_{A'} = \dfrac{0 + 0}{0 + 0 + u_A + v_A}$, $v = w = 0$. But then equation $y + v + w = \dfrac{x + y}{x + y + u + v}$, reduces to $y = y / (y + u)$, which implies that $y + u = 1$, which contradicts the stipulation that $u + v + x + y < 1$.

Cue Validity

Cue validity is the conditional probability that an object falls in a particular category given a particular feature or cue. The term was popularized by Beach, Reed and especially by Eleanor Rosch in her investigations of the acquisition of so-called basic categories.

Formally, the cue validity of a feature f_i with respect to category c_j has been defined in the following ways:

- As the conditional probability $p(c_j \mid f_i)$.
- As the deviation of the conditional probability from the category base rate, $p(c_j \mid f_i) - p(c_j)$.

- As a function of the linear correlation.
- Other definitions.

For the definitions based on probability, a high cue validity for a given feature means that the feature or attribute is more diagnostic of the class membership than a feature with low cue validity. Thus, a high-cue validity feature is one which conveys more information about the category or class variable, and may thus be considered as more useful for identifying objects as belonging to that category. Thus, high cue validity expresses high feature informativeness. For the definitions based on linear correlation, the expression of "informativeness" captured by the cue validity measure is not the full expression of the feature's informativeness (as in mutual information, for example), but only that portion of its informativeness that is expressed in a linear relationship. For some purposes, a bilateral measure such as the mutual information or category utility is more appropriate than the cue validity.

As an example, consider the domain of "numbers" and allow that every number has an attribute (i.e., a *cue*) named "is_positive_integer", which we call $f_{p\text{-}int,}$ and which adopts the value 1 if the number is actually a positive integer. Then we can inquire what the validity of this cue is with regard to the following classes: {rational number, irrational number, even integer}:

- If we know that a number is a positive integer we know that it is a rational number. Thus, $p(c_{rational} \mid f_{p\text{-}int}) = 1$, the cue validity for is_positive_integer as a cue for the category rational number is 1.
- If we know that a number is a positive integer then we know that it is not an irrational number. Thus, $p(c_{irrational} \mid f_{p\text{-}int}) = 1$, the cue validity for is_positive_integer as a cue for the category irrational number is 0.
- If we know only that a number is a positive integer, then its chances of being even or odd are 50-50 (there being the same number of even and odd integers). Thus, $p(c_{even} \mid f_{point}) = 0.5$, the cue validity for is_positive_integer as a cue for the category even integer is 0.5, meaning that the attribute is_positive_integer is entirely uninformative about the number's membership in the class even integer.

In perception, "cue validity" is often short for ecological validity of a perceptual cue, and is defined as a correlation rather than a probability. In this definition, an uninformative perceptual cue has an ecological validity of 0 rather than 0.5.

Use of the Cue Validity

In much of the work on modeling human category learning, there has been the assumption made (and sometimes validated) that attentional weighting tracks the cue validity, or tracks some related measure of feature informativeness. This would imply that attributes are differently weighted by the perceptual system; informative or high-cue validity attributes being weighted more heavily, while uninformative or low-cue validity attributes are weighted more lightly or ignored altogether.

Conditional Variance

In probability theory and statistics, a conditional variance is the variance of a random variable given the value(s) of one or more other variables. Particularly in econometrics, the conditional variance is also known as the scedastic function or skedastic function. Conditional variances are important parts of autoregressive conditional heteroskedasticity (ARCH) models.

The conditional variance of a random variable *Y* given another random variable *X* is,

$$\mathrm{Var}(Y \mid X) = E\left(\left(Y - E(Y \mid X)\right)^2 \mid X\right)$$

The conditional variance tells us how much variance is left if we use $E(Y \mid X)$ to "predict" *Y*. Here, as usual, $E(Y \mid X)$ stands for the conditional expectation of *Y* given *X*, which we may recall, is a random variable itself (a function of *X*, determined up to probability one). As a result, $\mathrm{Var}(Y \mid X)$ itself is a random variable (and is a function of *X*).

Explanation and Relation to Least-squares

Recall that variance is the expected squared deviation between a random variable (say, Y) and its expected value. The expected value can be thought of as a reasonable prediction of the outcomes of the random experiment (in particular, the expected value is the best constant prediction when predictions are assessed by expected squared prediction error). Thus, one interpretation of variance is that it gives the smallest possible expected squared prediction error. If we have the knowledge of another random variable (X) that we can use to predict Y, we can potentially use this knowledge to reduce the expected squared error. As it turns out, the best prediction of Y given X is the conditional expectation. In particular, for any $f: \mathbb{R} \to \mathbb{R}$ measurable,

$$\begin{aligned} E[(Y - f(X))^2] &= E[(Y - E(Y \mid X) + E(Y \mid X) - f(X))^2] \\ &= E[E\{(Y - E(Y \mid X) + E(Y \mid X) - f(X))^2 \mid X\}] \\ &= E[\mathrm{Var}(Y \mid X)] + E[(E(Y \mid X) - f(X))^2]. \end{aligned}$$

By selecting $f(X) = E(Y \mid X)$, the second, nonnegative term becomes zero, showing the claim. Here, the second equality used the law of total expectation. We also see that the expected conditional variance of Y given X shows up as the irreducible error of predicting Y given only the knowledge of X.

Special Cases and Variations

Conditioning on Discrete Random Variables

When *X* takes on countable many values $S = \{x_1, x_1, \ldots\}$ with positive probability, i.e., it is a discrete random variable, we can introduce $\text{Var}(Y \mid X = x)$, the conditional variance of *Y* given that *X=x* for any *x* from *S* as follows:

$$\text{Var}(Y \mid X = x) = E((Y - E(Y \mid X = x))^2 \mid X = x),$$

where recall that $E(Z \mid X = x)$ is the conditional expectation of *Z* given that *X=x*, which is well-defined for $x \in S$. An alternative notation for $\text{Var}(Y \mid X = x)$ is $\text{Var}_{Y|X}(Y \mid x)$.

Here $\text{Var}(Y \mid X = x)$ defines a constant for possible values of *x*, and in particular, $\text{Var}(Y \mid X = x)$, is *not* a random variable.

The connection of this definition to $\text{Var}(Y \mid X)$ is as follows: Let S be as above and define the function $v : S \to \mathbb{R}$ as $v(x) = \text{Var}(Y \mid X = x)$. Then, almost surely.

Definition using Conditional Distributions

The "conditional expectation of *Y* given *X=x*" can also be defined more generally using the conditional distribution of *Y* given *X* (this exists in this case, as both here *X* and *Y* are real-valued).

In particular, letting $P_{Y|X}$ be the (regular) conditional distribution $P_{Y|X}$ of *Y* given *X*, i.e., $P_{Y|X} : \mathcal{B} \times \mathbb{R} \to [0,1]$ (the intention is that $P_{Y|X}(U, x) = P(Y \in U \mid X = x)$ almost surely over the support of *X*), we can define,

$$\text{Var}(Y \mid X = x) = \int (y - \int y' P_{Y|X}(dy' \mid x))^2 P_{Y|X}(dy \mid x).$$

This can, of course, be specialized to when Y is discrete itself (replacing the integrals with sums), and also when the conditional density of Y given X=x with respect to some underlying distribution exists.

Components of Variance

The law of total variance says,

$$\text{Var}(Y) = E(\text{Var}(Y \mid X)) + \text{Var}(E(Y \mid X)).$$

In words: the variance of *Y* is the sum of the expected conditional variance *Y* given *X* and the variance of the conditional expectation of *Y* given *X*. The first term captures the variation left after "using *X* to predict *Y*", while the second term captures the variation due to the mean of the prediction of *Y* due to the randomness of *X*.

Conditional Variance as a Random Variable

As with $E(Y|X)$, we can consider $Var(Y|X)$ as a random variable. For example, if Y = height and X = sex for persons in a certain population, then Var(height | sex) is the variable which assigns to each person in the population the variance of height for that person's sex.

Expected Value of the Conditional Variance: Since $Var(Y|X)$ is a random variable, we can talk about its expected value. Using the formula $Var(Y|X)=E(Y^2|X)-[E(Y|X)]^2$, we have:

$$E(Var(Y|X))=E(E(Y^2|X))-E([E(Y|X)]^2)$$

We have already seen that the expected value of the conditional expectation of a random variable is the expected value of the original random variable, so applying this to Y^2 gives:

$$E(Var(Y|X))=E(Y^2)-E([E(Y|X)]^2)$$

Variance of the Conditional Expected Value: For what comes next, we will need to consider the variance of the conditional expected value. Using the second formula for variance, we have:

$$Var(E(Y|X))=E([E(Y|X)]^2)-[E(E(Y|X))]^2$$

Since $E(E(Y|X))=E(Y)$, this gives:

$$Var(E(Y|X))=E([E(Y|X)]^2)-[E(Y)]^2.$$

Putting It Together: Note that equation $E(Var(Y|X))=E(Y^2)-E([E(Y|X)]^2)$ and,

$$Var(E(Y|X))=E([E(Y|X)]^2)-[E(Y)]^2$$

both contain the term $E([E(Y|X)]^2)$ but with opposite signs. So adding them gives:

$$E(Var(Y|X))+Var(E(Y|X))=E(Y^2)-[E(Y)]^2,$$

which is just $\text{Var}(Y)$. In other words,

$$\text{Var}(Y) = E(\text{Var}(Y|X)) + \text{Var}(E(Y|X)).$$

In words: The marginal variance is the sum of the expected value of the conditional variance and the variance of the conditional means.

Consequences:

This says that two things contribute to the marginal (overall) variance: the expected value of the conditional variance, and the variance of the conditional means. Moreover, $\text{Var}(Y) = E(\text{Var}(Y|X))$ if and only if $\text{Var}(E(Y|X)) = 0$. What would this say about $E(Y|X)$?

Since variances are always non-negative, equation $\text{Var}(Y) = E(\text{Var}(Y|X)) + \text{Var}(E(Y|X))$ implies:

$$\text{Var}(Y) \geq E(\text{Var}(Y|X)).$$

Since $\text{Var}(Y|X) \geq 0$, $E(\text{Var}(Y|X))$ must also be ≥ 0. equation:

$$\text{Var}(Y) = E(\text{Var}(Y|X)) + \text{Var}(E(Y|X))$$

implies, $\text{Var}(Y) \geq \text{Var}(E(Y|X))$.

Moreover, $\text{Var}(Y) = \text{Var}(E(Y|X))$ if and only if $E(\text{Var}(Y|X)) = 0$. What would this imply about $\text{Var}(Y|X)$ and about the relationship between Y and X?

Another perspective on equation $\text{Var}(Y) = E(\text{Var}(Y|X)) + \text{Var}(E(Y|X))$ i) E($\text{Var}(Y|X)$ is a weighted average of $\text{Var}(Y|X)$.

$$\begin{aligned}\text{Var}(E(Y|X) &= E\left(\left[E(Y|X) - E(E(Y|X))\right]^2\right) \\ &= E\left(\left[E(Y|X) - (E(y))\right]^2\right),\end{aligned}$$

which is a weighted average of $[E(Y|X) - (E(Y)]^2$.

Thus, equation $\text{Var}(Y) = E(\text{Var}(Y|X)) + \text{Var}(E(Y|X))$ says that $\text{Var}(Y)$ is a weighted mean of $\text{Var}(Y|X)$ plus a weighted mean of $[E(Y|X) - (E(Y)]^2$ (and is a weighted mean of $\text{Var}(Y|X)$ if and only if all conditional expected values $E(Y|X)$ are equal to the marginal expected value $E(Y)$.)

References

- Draheim, Dirk (2017). "An Operational Semantics of Conditional Probabilities that Fully Adheres to Kolmogorov's Explication of Probability Theory". doi:10.13140/RG.2.2.10050.48323/3
- Grimmett, Geoffrey; Stirzaker, David (2001). Probability and Random Processes (3rd ed.). Oxford University Press. ISBN 0-19-857222-0
- Conditional-probability, other, knowledge, resources: corporatefinanceinstitute.com, Retrieved 21 May, 2020
- Draheim, Dirk (2017). "Generalized Jeffrey Conditionalization (A Frequentist Semantics of Partial Conditionalization)". Springer. Retrieved December 19, 2017

Probability Distributions

4

- Random Variable
- Probability Distribution Function
- Discrete Distributions
- Continuous Distributions

A mathematical and statistical function that describes all the possible values and likelihood that a random variable can have is called probability distribution. Cumulative probability distribution, discrete probability distribution, continuous probability distribution, etc. are studied under its domain. This chapter discusses the subject of probability distribution in detail.

Random Variable

A variable is something which can change its value. It may vary with different outcomes of an experiment. If the value of a variable depends upon the outcome of a random experiment it is a random variable. A random variable can take up any real value.

Mathematically, a random variable is a real-valued function whose domain is a sample space S of a random experiment. A random variable is always denoted by capital letter like X, Y, M etc. The lowercase letters like x, y, z, m etc. represent the value of the random variable.

Consider the random experiment of tossing a coin 20 times. You will earn \$5 is you get head and will lose \$5 if it a tail. You and your friend are all set to see who will win the game by earning more money. Here, we see that the value of getting head for the coin tossed for 20 times is anything from zero to twenty. If we denote the number of a head by X, then $X = \{0, 1, 2, \ldots, 20\}$. The probability of getting a head is always ½.

Properties of a Random Variable

- It only takes the real value.
- If X is a random variable and C is a constant, then CX is also a random variable.
- If X_1 and X_2 are two random variables, then $X_1 + X_2$ and $X_1 X_2$ are also random.
- For any constants C_1 and C_2, $C_1X_1 + C_2X_2$ is also random.
- $|X|$ is a random variable.

Types of Random Variable

Discrete Random Variables

A discrete random variable is one which may take on only a countable number of distinct values such as 0, 1, 2, 3, 4,…,. Discrete random variables are usually (but not necessarily) counts. If a random variable can take only a finite number of distinct values, then it must be discrete. Examples of discrete random variables include the number of children in a family, the Friday night attendance at a cinema, the number of patients in a doctor's surgery, the number of defective light bulbs in a box of ten.

The probability distribution of a discrete random variable is a list of probabilities associated with each of its possible values. It is also sometimes called the probability function or the probability mass function.

Suppose a random variable X may take k different values, with the probability that $X = x_i$ defined to be $P(X = x_i) = p_i$. The probabilities pi must satisfy the following:

$$0 < p_i < 1 \text{ for each } i$$

$$p_1 + p_2 + \ldots + p_k = 1.$$

Example: Suppose a variable X can take the values 1, 2, 3, or 4. The probabilities associated with each outcome are described by the following table:

Outcome	1	2	3	4
Probability	0.1	0.3	0.4	0.2

The probability that X is equal to 2 or 3 is the sum of the two probabilities: $P(X = 2 \text{ or } X = 3) = P(X = 2) + P(X = 3) = 0.3 + 0.4 = 0.7$. Similarly, the probability that X is greater than 1 is equal to $1 - P(X = 1) = 1 - 0.1 = 0.9$, by the complement rule.

All random variables (discrete and continuous) have a cumulative distribution function. It is a function giving the probability that the random variable X is less than or equal to x, for every value x. For a discrete random variable, the cumulative distribution function is found by summing up the probabilities.

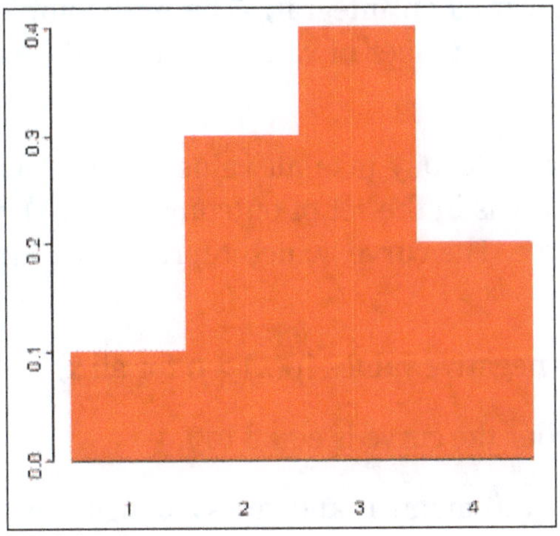

Probability Histogram.

The cumulative distribution function for the above probability distribution is calculated as follows:

- The probability that X is less than or equal to 1 is 0.1.
- The probability that X is less than or equal to 2 is 0.1+0.3 = 0.4.
- The probability that X is less than or equal to 3 is 0.1+0.3+0.4 = 0.8.
- The probability that X is less than or equal to 4 is 0.1+0.3+0.4+0.2 = 1.

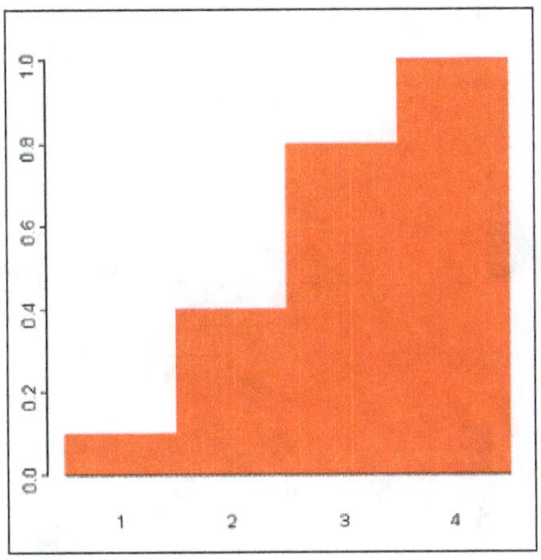

Continuous Random Variables

A continuous random variable is one which takes an infinite number of possible values. Continuous random variables are usually measurements. Examples include height, weight, the amount of sugar in an orange, the time required to run a mile.

A continuous random variable is not defined at specific values. Instead, it is defined over an interval of values, and is represented by the area under a curve (in advanced mathematics, this is known as an integral). The probability of observing any single value is equal to 0, since the number of values which may be assumed by the random variable is infinite.

Suppose a random variable X may take all values over an interval of real numbers. Then the probability that X is in the set of outcomes A, P(A), is defined to be the area above A and under a curve. The curve, which represents a function p(x), must satisfy the following:

- The curve has no negative values (p(x) > 0 for all x).
- The total area under the curve is equal to 1.

A curve meeting these requirements is known as a density curve.

The Uniform Distribution

A random number generator acting over an interval of numbers (a, b) has a continuous distribution. Since any interval of numbers of equal width has an equal probability of being observed, the curve describing the distribution is a rectangle, with constant height across the interval and 0 height elsewhere. Since the area under the curve must be equal to 1, the length of the interval determines the height of the curve.

The following graphs plot the density curves for random number generators over the intervals (4,5) (top left), (2,6) (top right), (5,5.5) (lower left), and (3,5) (lower right). The distributions corresponding to these curves are known as uniform distributions.

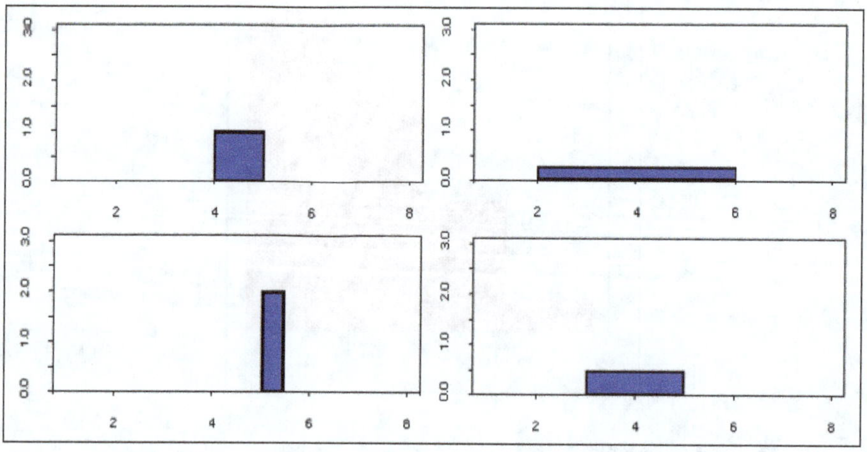

Consider the uniform random variable X defined on the interval (2,6). Since the interval has width = 4, the curve has height = 0.25 over the interval and 0 elsewhere. The probability that X is less than or equal to 5 is the area between 2 and 5, or (5-2)*0.25 = 0.75. The probability that X is greater than 3 but less than 4 is the area between 3 and

4, (4-3)*0.25 = 0.25. To find that probability that X is less than 3 or greater than 5, add the two probabilities:

$$P(X \le 3 \text{ and } X \ge 5) = P(X \le 3) + P(X \ge 5) = (3-2)*0.25 + (6-5)*0.25 = 0.25 + 0.25 = 0.5.$$

The uniform distribution is often used to simulate data. Suppose you would like to simulate data for 10 rolls of a regular 6-sided die. Using the MINITAB "RAND" command with the "UNIF" subcommand generates 10 numbers in the interval (0,6):

```
MTB > RAND 10 c2;

SUBC> unif 0 6.
```

Assign the discrete random variable X to the values 1, 2, 3, 4, 5, or 6 as follows:

if $0 < X < 1,\ X = 1$

if $1 < X < 2,\ X = 2$

if $2 < X < 3,\ X = 3$

if $3 < X < 4,\ X = 4$

if $4 < X < 5,\ X = 5$

if $X > 5,\ X = 6.$

Use the generated MINITAB data to assign X to a value for each roll of the die:

Uniform Data	X Value
4.53786	5
5.77474	6
3.69518	4
1.03929	2
4.23835	5
0.37096	1
0.75272	1
5.56563	6
0.89045	1
3.18086	4

Another type of continuous density curve is the normal distribution. The area under the curve is not easy to calculate for a normal random variable X with mean μ and standard deviation σ. However, tables (and computer functions) are available for the standard random variable Z, which is computed from X by subtracting μ and dividing by σ. All of the rules of probability apply to the normal distribution.

Probability Distribution Function

A probability distribution function is some function that may be used to define a particular probability distribution. Depending upon which text is consulted.

Cumulative Distribution Function

In probability theory and statistics, the cumulative distribution function (CDF) of a real-valued random variable X, or just distribution function of X, evaluated at x, is the probability that X will take a value less than or equal to x.

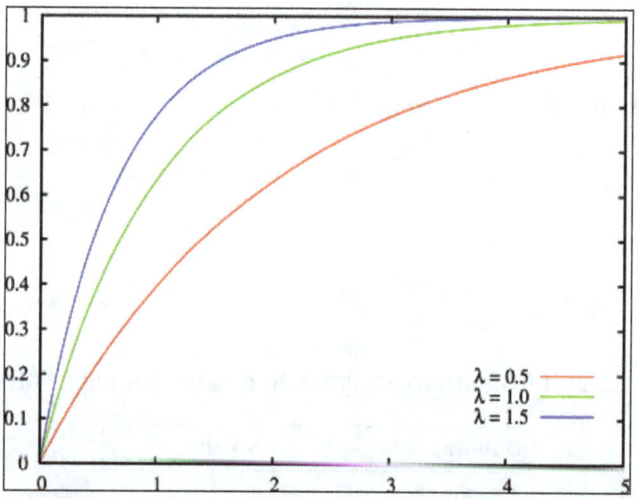

Cumulative distribution function for the exponential distribution.

In the case of a scalar continuous distribution, it gives the area under the probability density function from minus infinity to x. Cumulative distribution functions are also used to specify the distribution of multivariate random variables.

The cumulative distribution function of a real-valued random variable X is the function given by:

$$F_X(x) = P(X \leq x)$$

where the right-hand side represents the probability that the random variable X takes on a value less than or equal to x. The probability that X lies in the semi-closed interval $(a, b]$, where $a < b$, is therefore,

$$P(a < X \leq b) = F_X(b) - F_X(a)$$

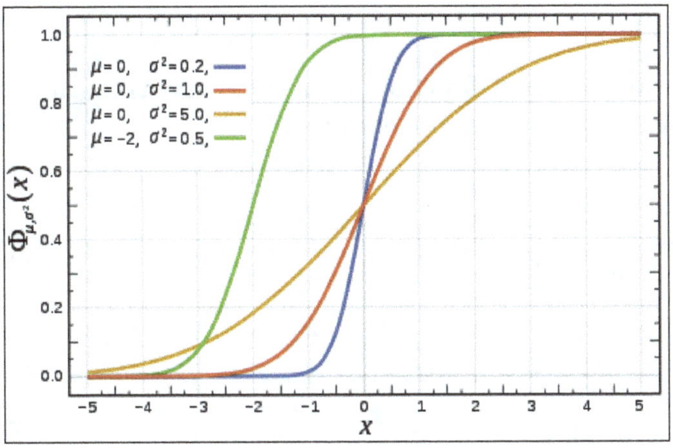

Cumulative distribution function for the normal distribution.

In the definition above, the "less than or equal to" sign, "≤", is a convention, not a universally used one (e.g. Hungarian literature uses "<"), but the distinction is important for discrete distributions. The proper use of tables of the binomial and Poisson distributions depends upon this convention. Moreover, important formulas like Paul Lévy's inversion formula for the characteristic function also rely on the "less than or equal" formulation.

If treating several random variables $X, Y, \ldots$ etc. the corresponding letters are used as subscripts while, if treating only one, the subscript is usually omitted. It is conventional to use a capital F for a cumulative distribution function, in contrast to the lower-case f used for probability density functions and probability mass functions. This applies when discussing general distributions: Some specific distributions have their own conventional notation, for example the normal distribution.

The probability density function of a continuous random variable can be determined from the cumulative distribution function by differentiating.

Probability density function from the cumulative distribution function,

Given $F(x)$,

$$f(x) = \frac{dF(x)}{dx}, \text{ as long as the derivative exists.}$$

The CDF of a continuous random variable X can be expressed as the integral of its probability density function f_X as follows:

$$F_X(x) = \int_{-\infty}^{x} f_X(t)\,dt.$$

In the case of a random variable X which has distribution having a discrete component at a value b,

$$P(X = b) = F_X(b) - \lim_{x \to b^-} F_X(x).$$

If F_x is continuous at b, this equals zero and there is no discrete component at b.

Properties

Every cumulative distribution function F_x is non-decreasing and right-continuous, which makes it a càdlàg function. Furthermore,

$$\lim_{x\to-\infty} F_X(x) = 0, \quad \lim_{x\to+\infty} F_X(x) = 1.$$

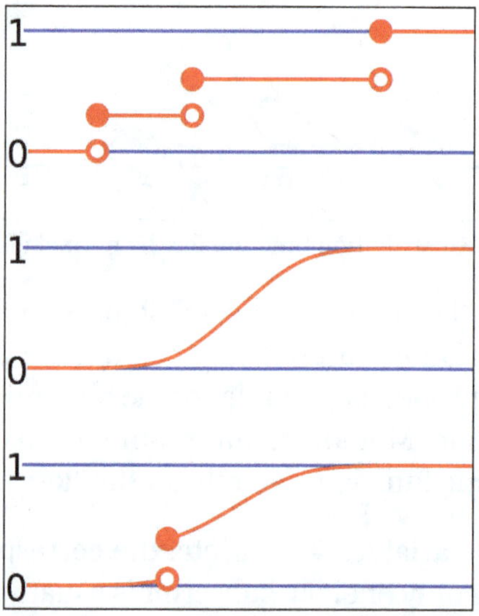

From top to bottom, the cumulative distribution function of a discrete probability distribution, continuous probability distribution, and a distribution which has both a continuous part and a discrete part.

Every function with these four properties is a CDF, i.e., for every such function, a random variable can be defined such that the function is the cumulative distribution function of that random variable.

If X is a purely discrete random variable, then it attains values $x_1, x_2, \ldots$ with probability $p_i = p(x_i)$, and the CDF of X will be discontinuous at the points x_i:

$$F_X(x) = P(X \le x) = \sum_{x_i \le x} P(X = x_i) = \sum_{x_i \le x} p(x_i).$$

If the CDF F_x of a real valued random variable X is continuous, then X is a continuous random variable; if furthermore F_x is absolutely continuous, then there exists a Lebesgue-integrable function $f_X(x)$ such that,

$$F_X(b) - F_X(a) = P(a < X \le b) = \int_a^b f_X(x)dx$$

for all real numbers a and b. The function f_x is equal to the derivative of F_x almost everywhere, and it is called the probability density function of the distribution of X.

As an example, suppose X is uniformly distributed on the unit interval [0,1].

Then the CDF of X is given by,

$$F_X(x) = \begin{cases} 0 & : x < 0 \\ x & : 0 \le x \le 1 \\ 1 & : x > 1 \end{cases}$$

Suppose instead that X takes only the discrete values 0 and 1, with equal probability.

Then the CDF of X is given by,

$$F_X(x) = \begin{cases} 0 & : x < 0 \\ 1/2 & : 0 \le x < 1 \\ 1 & : x \ge 1 \end{cases}$$

Suppose X is exponential distributed. Then the CDF of X is given by,

$$F_X(x;\lambda) = \begin{cases} 1 - e^{-\lambda x} & x \ge 0, \\ 0 & x < 0. \end{cases}$$

Here $\lambda > 0$ is the parameter of the distribution, often called the rate parameter.

Suppose X is normal distributed. Then the CDF of X is given by,

$$F(x;\mu,\sigma) = \frac{1}{\sigma\sqrt{2\pi}} \int_{-\infty}^{x} \exp\left(-\frac{(t-\mu)^2}{2\sigma^2} \right) dt.$$

Here the parameter μ is the mean or expectation of the distribution; and σ is its standard deviation.

Suppose X is binomial distributed. Then the CDF of X is given by,

$$F(k;n,p) = \Pr(X \le k) = \sum_{i=0}^{\lfloor k \rfloor} \binom{n}{i} p^i (1-p)^{n-i}$$

Here p is the probability of success and the function denotes the discrete probability distribution of the number of successes in a sequence of n independent experiments, and $\lfloor k \rfloor$ is the "floor" under k, i.e. the greatest integer less than or equal to k.

Derived Functions

Complementary Cumulative Distribution Function (Tail Distribution)

Sometimes, it is useful to study the opposite question and ask how often the random variable is above a particular level. This is called the complementary cumulative distribution function (ccdf) or simply the tail distribution or exceedance, and is defined as,

$$\bar{F}_X(x) = P(X > x) = 1 - F_X(x).$$

This has applications in statistical hypothesis testing, for example, because the one-sided p-value is the probability of observing a test statistic *at least* as extreme as the one observed. Thus, provided that the test statistic, *T*, has a continuous distribution, the one-sided p-value is simply given by the ccdf: For an observed value t of the test statistic,

$$p = P(T \geq t) = P(T > t) = 1 - F_T(t).$$

In survival analysis, $\bar{F}_X(x)$ is called the survival function and denoted $S(x)$, while the term *reliability function* is common in engineering.

Z-table: One of the most popular application of cumulative distribution function is standard normal table, also called the unit normal table or Z table, is the value of cumulative distribution function of the normal distribution. It is very useful to use Z-table not only for probabilities below a value which is the original application of cumulative distribution function, but also above and/or between values on standard normal distribution, and it was further extended to any normal distribution.

Properties:

- For a non-negative continuous random variable having an expectation, Markov's inequality states that,

  $$\bar{F}_X(x) \leq \frac{E(X)}{x}.$$

- As $x \to \infty, \bar{F}_X(x) \to 0$, and in fact $\bar{F}_X(x) = o(1/x)$ provided that $E(X)$ is finite.

  Proof: Assuming X has a density function f_X, for any $c > 0$,

  $$E(X) = \int_0^\infty x f_X(x)dx \geq \int_0^c x f_X(x)dx + c\int_0^\infty f_x(x)dx$$

  Then, on recognizing $\bar{F}_X(c) = \int_c^\infty f_X(x)dx$ and rearranging terms,

  $$0 \leq c\bar{F}_X(c) \leq E(X) - \int_0^c x f_X(x)dx \to 0 \text{ as } c \to \infty$$ as claimed.

Folded Cumulative Distribution

While the plot of a cumulative distribution often has an S-like shape, an alternative illustration is the folded cumulative distribution or mountain plot, which folds the top half of the graph over, thus using two scales, one for the upslope and another for the downslope. This form of illustration emphasises the median and dispersion (specifically, the mean absolute deviation from the median) of the distribution or of the empirical results.

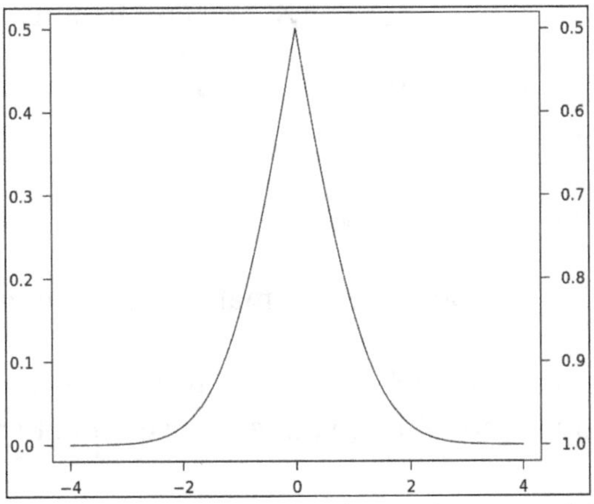

Example of the folded cumulative distribution for a normal distribution function with an expected value of 0 and a standard deviation of 1.

Inverse Distribution Function (Quantile Function)

If the CDF *F* is strictly increasing and continuous then $F^{-1}(p), p \in [0,1]$, is the unique real number x such that $F(x) = p$. In such a case, this defines the inverse distribution function or quantile function.

Some distributions do not have a unique inverse (for example in the case where $f_X(x) = 0$ for all $a < x < b$, causing F_X to be constant). This problem can be solved by defining, for $p \in [0,1]$, the generalized inverse distribution function:

$$F^{-1}(p) = \inf\{x \in \mathbb{R} : F(x) \geq p\}.$$

- Example: The median is $F^{-1}(0.5)$.
- Example: Put $\tau = F^{-1}(0.95)$. Then we call τ the 95th percentile.

Some useful properties of the inverse cdf (which are also preserved in the definition of the generalized inverse distribution function) are:

- F^{-1} is nondecreasing.
- $F^{-1}(F(x)) \leq x$
- $F(F^{-1}(p)) \geq p$
- $F^{-1}(p) \leq x$ if and only if $p \leq F(x)$
- If Y has a $U[0,1]$ distribution then $F^{-1}(Y)$ is distributed as F. This is used in random number generation using the inverse transform sampling-method.

- If $\{X_\alpha\}$ is a collection of independent F-distributed random variables defined on the same sample space, then there exist random variables Y_α such that Y_α is distributed as $U[0,1]$ and $F^{-1}(Y_\alpha) = X_\alpha$ with probability 1 for all α.

The inverse of the cdf can be used to translate results obtained for the uniform distribution to other distributions.

Empirical Distribution Function

The empirical distribution function is an estimate of the cumulative distribution function that generated the points in the sample. It converges with probability 1 to that underlying distribution. A number of results exist to quantify the rate of convergence of the empirical distribution function to the underlying cumulative distribution function.

Multivariate Case

Definition for Two Random Variables

When dealing simultaneously with more than one random variable the joint cumulative distribution function can also be defined. For example, for a pair of random variables X, Y, the joint CDF F_{XY} is given by:

$$F_{XY}(x,y) = P(X \leq x, Y \leq y)$$

where the right-hand side represents the probability that the random variable X takes on a value less than or equal to x and that Y takes on a value less than or equal to y.

Example of joint cumulative distribution function:

For two continuous variables X and Y: $\Pr(a < X < b \text{ and } c < Y < d) = \int_a^b \int_c^d f(x,y)\,dy\,dx$;

For two discrete random variables, it is beneficial to generate a table of probabilities and address the cumulative probability for each potential range of *X* and *Y*, and here is the example:

Given the joint probability density function in tabular form, determine the joint cumulative distribution function.

	Y = 2	Y = 4	Y = 6	Y = 8
X = 1	0	0.1	0	0.1
X = 3	0	0	0.2	0
X = 5	0.3	0	0	0.15
X = 7	0	0	0.15	0

Using the given table of probabilities for each potential range of X and Y, the joint cumulative distribution function may be constructed in tabular form:

	$Y < 2$	$2 \le Y < 4$	$4 \le Y < 6$	$6 \le Y < 8$	$Y \le 8$
$X < 1$	0	0	0	0	0
$1 \le X < 3$	0	0	0.1	0.1	0.2
$3 \le X < 5$	0	0	0.1	0.3	0.4
$5 \le X < 7$	0	0.3	0.4	0.6	0.85
$X \le 7$	0	0.3	0.4	0.75	1

Definition for More than Two Random Variables

For N random variables $X_1,\ldots,X_N$, the joint CDF $F_{X_1,\ldots,X_N}$ is given by,

$$F_{X_1,\ldots,X_N}(x_1,\ldots,x_N) = P(X_1 \le x_1,\ldots,X_N \le x_n)$$

Interpreting the N random variables as a random vector $X = (X_1,\ldots,X_N)^T$ yields a shorter notation:

$$F_X(x) = P(X_1 \le x_1,\ldots,X_N \le x_n)$$

Properties

Every multivariate CDF is:

- Monotonically non-decreasing for each of its variables,
- Right-continuous in each of its variables,

  $0 \le F_{X_1 \ldots X_n}(x_1,\ldots,x_n) \le 1,$

  $\lim_{x_1,\ldots,x_n \to +\infty} F_{X_1 \ldots X_n}(x_1,\ldots,x_n) = 1$ and $\lim_{x_i \to -\infty} F_{X_1 \ldots X_n}(x_1,\ldots,x_n) = 0,$ for all i.

Complex Case

Complex Random Variable

The generalization of the cumulative distribution function from real to complex random variables is not obvious because expressions of the form $P(Z \le 1+2i)$ make no sense. However expressions of the form $P(\Re(Z) \le 1, \Im(Z) \le 3)$ make sense. Therefore, we define the cumulative distribution of a complex random variables via the joint distribution of their real and imaginary parts:

$$F_Z(z) = F_{\Re(Z),\Im(Z)}(\Re(z),\Im(z)) = P(\Re(Z) \le \Re(z), \Im(Z) \le \Im(z)).$$

Complex Random Vector

Generalization of equation $F_{X_1,\ldots,X_N}(x_1,\ldots,x_N) = P(X_1 \le x_1,\ldots,X_N \le x_n)$ yields,

$$F_Z(z) = F_{\Re(Z_1),\Im(Z_1),\ldots,\Re(Z_n),\Im(Z_n)}(\Re(z_1),\Im(z_1),\ldots,$$
$$\Re(z_n),\Im(z_n)) = P(\Re(Z_1) \le \Re(z_1),\Im(Z_1) \le \Im(z_1),\ldots,\Re(Z_n) \le \Re(z_n),\Im(Z_n) \le \Im(z_n))$$

as definition for the CDS of a complex random vector $Z = (Z_1,\ldots,Z_N)^T$.

Use in Statistical Analysis

The concept of the cumulative distribution function makes an explicit appearance in statistical analysis in two (similar) ways. Cumulative frequency analysis is the analysis of the frequency of occurrence of values of a phenomenon less than a reference value. The empirical distribution function is a formal direct estimate of the cumulative distribution function for which simple statistical properties can be derived and which can form the basis of various statistical hypothesis tests. Such tests can assess whether there is evidence against a sample of data having arisen from a given distribution, or evidence against two samples of data having arisen from the same (unknown) population distribution.

Kolmogorov–Smirnov and Kuiper's Tests

The Kolmogorov–Smirnov test is based on cumulative distribution functions and can be used to test to see whether two empirical distributions are different or whether an empirical distribution is different from an ideal distribution. The closely related Kuiper's test is useful if the domain of the distribution is cyclic as in day of the week. For instance Kuiper's test might be used to see if the number of tornadoes varies during the year or if sales of a product vary by day of the week or day of the month.

Probability Mass Function

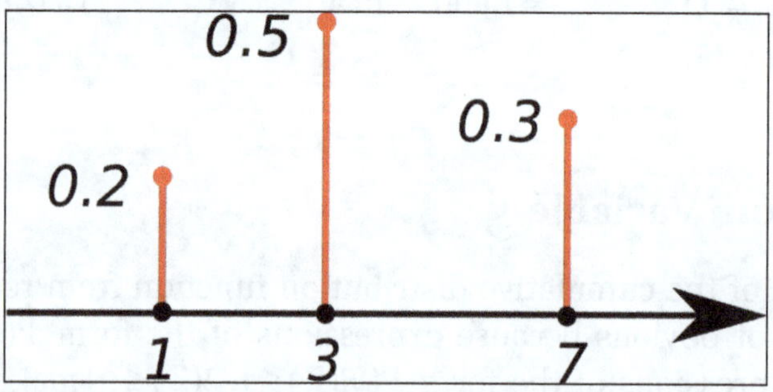

The graph of a probability mass function. All the values of this function must be non-negative and sum up to 1.

In probability and statistics, a probability mass function (PMF) is a function that gives the probability that a discrete random variable is exactly equal to some value.

Sometimes it is also known as the discrete density function. The probability mass function is often the primary means of defining a discrete probability distribution, and such functions exist for either scalar or multivariate random variables whose domain is discrete.

A probability mass function differs from a probability density function (PDF) in that the latter is associated with continuous rather than discrete random variables. A PDF must be integrated over an interval to yield a probability. The value of the random variable having the largest probability mass is called the mode.

Probability mass function is the probability distribution of a discrete random variable, and provides the possible values and their associated probabilities. It is the function p: $\mathbb{R}$ [0,1] defined by,

$$p_X(x_i) = P(X = x_i)$$

for $-\infty < x < \infty$. $p_X(x)$ can also be simplified as $p_X(x)$.

The probabilities associated with each possible values must be positive and sum up to for all other values, the probabilities need to be 0.

$$\sum p_X(x_i) = 1$$

$$p(x_i) > 0$$

$$p(x) = 0 \text{ for all other } x$$

Thinking of probability as mass helps to avoid mistakes since the physical mass is conserved as is the total probability for all hypothetical outcomes x.

Measure Theoretic Formulation

A probability mass function of a discrete random variable X can be seen as a special case of two more general measure theoretic constructions: The distribution of X and the probability density function of X with respect to the counting measure.

Suppose that $(A, \mathcal{A}, P)$ is a probability space and that $(B, \mathcal{B})$ is a measurable space whose underlying σ-algebra is discrete, so in particular contains singleton sets of B. In this setting, a random variable $X : A \to B$ is discrete provided its image is countable. The pushforward measure $X_*(P)$ —called a distribution of X in this context—is a probability measure on B whose restriction to singleton sets induces a probability mass function $f_X : B \to \mathbb{R}$ since $f_X(b) = P(X^{-1}(b)) = [X_*(P)](\{b\})$ for each $b \in B$.

Now suppose that $(B, \mathcal{B}, \mu)$ is a measure space equipped with the counting measure μ. The probability density function f of X with respect to the counting measure, if it exists, is the Radon–Nikodym derivative of the pushforward measure of X (with respect to the counting measure), so $f = dX_*P / d\mu$ and f is a function from B to the non-negative reals. As a consequence, for any $b \in B$ we have,

$$P(X = b) = P(X^{-1}(\{b\})) := \int_{X^{-1}(\{b\})} dP = \int_{\{b\}} f d\mu = f(b),$$

demonstrating that f is in fact a probability mass function.

When there is a natural order among the potential outcomes x, it may be convenient to assign numerical values to them (or *n*-tuples in case of a discrete multivariate random variable) and to consider also values not in the image of X. That is, f_X may be defined for all real numbers and $f_X(x) = 0$ for all $x \notin X(S)$ as shown in the figure.

The image of X has a countable subset on which the probability mass function $f_X(x)$ is one. Consequently, the probability mass function is zero for all but a countable number of values of x.

The discontinuity of probability mass functions is related to the fact that the cumulative distribution function of a discrete random variable is also discontinuous. If X is a discrete random variable, $P(X = x) = 1$ then means that the casual event $(X = x)$ is certain (it is true in the 100% of the occurrencies); on the contrary, $P(X = x) = 0$ means that the casual event $(X = x)$ is always impossible. This statement isn't true for a continuous random variable X, for which $P(X = x) = 0$ for any possible x: In fact, by definition, a continuous random variable can have an infinite set of possible values and thus the probability it has a single particular value *x* is equal to $\frac{1}{\infty} = 0$. Discretization is the process of converting a continuous random variable into a discrete one.

Finite

There are three major distributions associated, the Bernoulli distribution, the Binomial distribution and the geometric distribution:

- Bernoulli distribution, Ber(p), is used to model an experiment with only two possible outcomes. The two outcomes are often encoded as 1 and 0.

  $$p_X(x) = \begin{cases} p, & \text{if x is 1} \\ 1-p, & \text{if x is 0} \end{cases}$$

  An example of the Bernoulli distribution is tossing a coin. Suppose that S is the sample space of all outcomes of a single toss of a fair coin, and X is the random variable defined on S assigning 0 to the category "tails" and 1 to the category "heads". Since the coin is fair, the probability mass function.

  $$p_X(x) = \begin{cases} \frac{1}{2}, & x \in \{0,1\}, \\ 0, & x \notin \{0,1\}. \end{cases}$$

Binomial distribution, Bin(n,p), models the number of successes when someone draws n times with replacement. Each draw or experiment is independent, with two possible outcomes. The associated probability mass function is $\binom{n}{k}p^k(1-p)^{n-k}$. An example of the Binomial distribution is the probability of getting exactly one 6 when someone rolls a fair die three times:

- Geometric distribution describes the number of trials needed to get one success, denoted as Geo(p). Its probability mass function is $p_X(k) = (1-p)^{k-1}p$. An example is tossing the coin until the first head appears.

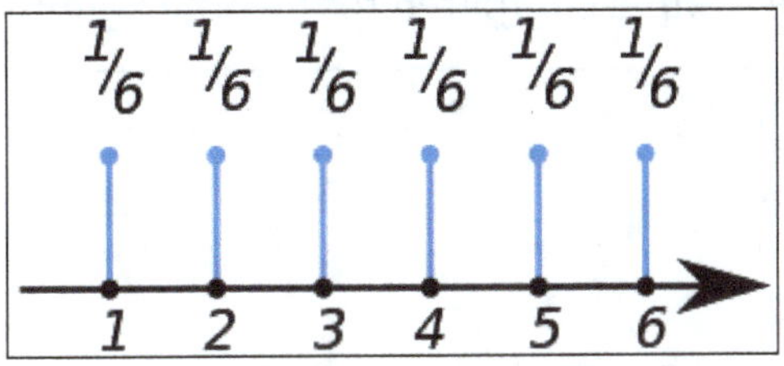

- The probability mass function of a fair die. All the numbers on the die have an equal chance of appearing on top when the die stops rolling.

Other distributions that can be modeled using a probability mass function is the Categorical distribution (also known as the generalized Bernoulli distribution) and the multinomial distribution:

- If the discrete distribution has two or more categories one of which may occur, whether or not these categories have a natural ordering, when there is only a single trial (draw) this is a categorical distribution.

- An example of a multivariate discrete distribution, and of its probability mass function, is provided by the multinomial distribution. Here the multiple random variables are the numbers of successes in each of the categories after a given number of trials, and each non-zero probability mass gives the probability of a certain combination of numbers of successes in the various categories.

Infinite

- The following exponentially declining distribution is an example of a distribution with an infinite number of possible outcomes—all the positive integers:

  $$\Pr(X = i) = \frac{1}{2^i} \quad \text{for} \quad i = 1, 2, 3, \ldots$$

  Despite the infinite number of possible outcomes, the total probability mass is 1/2 + 1/4 + 1/8 + ... = 1, satisfying the unit total probability requirement for a probability distribution.

Probability Density Function

In probability theory, a probability density function (PDF), or density of a continuous random variable, is a function whose value at any given sample (or point) in the sample space (the set of possible values taken by the random variable) can be interpreted as providing a relative likelihood that the value of the random variable would equal that sample. In other words, while the absolute likelihood for a continuous random variable to take on any particular value is 0 (since there are an infinite set of possible values to begin with), the value of the PDF at two different samples can be used to infer, in any particular draw of the random variable, how much more likely it is that the random variable would equal one sample compared to the other sample.

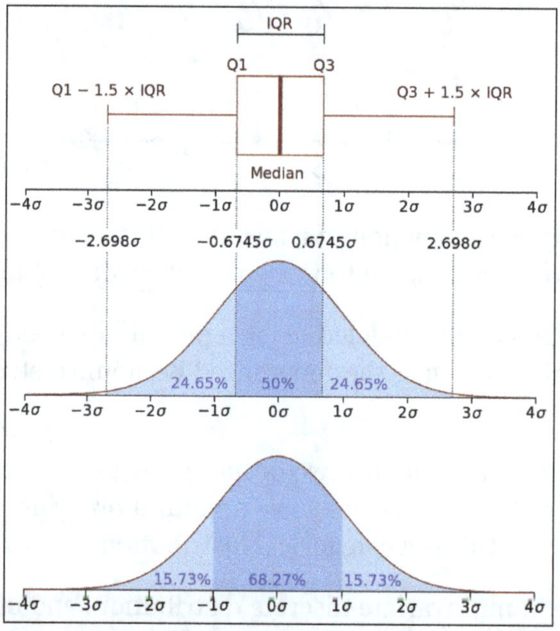

Boxplot and probability density function of a normal distribution $N(0, \sigma^2)$.

In a more precise sense, the PDF is used to specify the probability of the random variable falling within a particular range of values, as opposed to taking on any one value. This probability is given by the integral of this variable's PDF over that range—that is, it is given by the area under the density function but above the horizontal axis and between the lowest and greatest values of the range. The probability density function is nonnegative everywhere, and its integral over the entire space is equal to 1.

The terms "probability distribution function" and "probability function" have also sometimes been used to denote the probability density function. However, this use is not standard among probabilists and statisticians. In other sources, "probability distribution function" may be used when the probability distribution is defined as a function over general sets of values or it may refer to the cumulative distribution function, or it may be a probability mass function (PMF) rather than the density. "Density function" itself is also used for the probability mass function, leading to further confusion. In general though, the PMF is used in the context of discrete random variables (random variables that take values on a countable set), while the PDF is used in the context of continuous random variables.

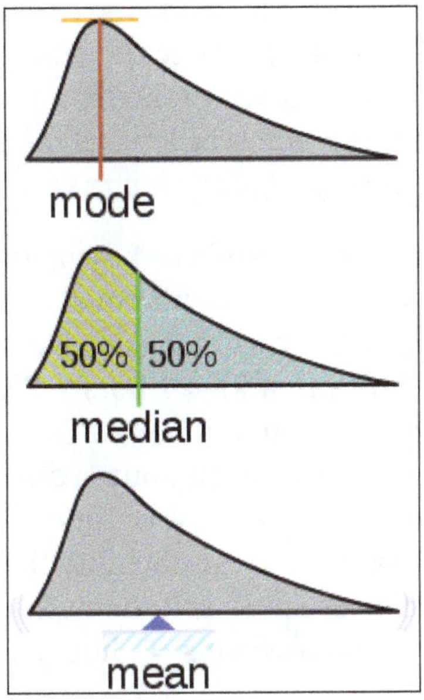

Geometric visualisation of the mode, median and mean of an arbitrary probability density function.

A random variable X with values in a measurable space $(\mathcal{X},\mathcal{A})$ (usually $\mathbb{R}^n$ with the Borel sets as measurable subsets) has as probability distribution the measure X_*P on $(\mathcal{X},\mathcal{A})$: The density of X with respect to a reference measure μ on $(\mathcal{X},\mathcal{A})$ is the Radon–Nikodym derivative:

$$f=\frac{dX_*P}{d\mu}.$$

That is, *f* is any measurable function with the property that:

$$\Pr[X\in A]=\int_{X^{-1}A}dP=\int_A f\,d\mu$$

for any measurable set $A\in\mathcal{A}$.

In the continuous univariate case above, the reference measure is the Lebesgue measure. The probability mass function of a discrete random variable is the density with respect to the counting measure over the sample space (usually the set of integers, or some subset thereof).

It is not possible to define a density with reference to an arbitrary measure (e.g. one can't choose the counting measure as a reference for a continuous random variable). Furthermore, when it does exist, the density is almost everywhere unique.

Example: Suppose bacteria of a certain species typically live 4 to 6 hours. The probability that a bacterium lives exactly 5 hours is equal to zero. A lot of bacteria live for approximately 5 hours, but there is no chance that any given bacterium dies at exactly 5.0000000000... hours. However, the probability that the bacterium dies between 5 hours and 5.01 hours is quantifiable. Suppose the answer is 0.02 (i.e., 2%). Then, the probability that the bacterium dies between 5 hours and 5.001 hours should be about 0.002, since this time interval is one-tenth as long as the previous. The probability that the bacterium dies between 5 hours and 5.0001 hours should be about 0.0002, and so on.

In these three examples, the ratio (probability of dying during an interval) / (duration of the interval) is approximately constant, and equal to 2 per hour (or 2 $hour^{-1}$). For example, there is 0.02 probability of dying in the 0.01-hour interval between 5 and 5.01 hours, and (0.02 probability / 0.01 hours) = 2 $hour^{-1}$. This quantity 2 $hour^{-1}$ is called the probability density for dying at around 5 hours. Therefore, the probability that the bacterium dies at 5 hours can be written as (2 $hour^{-1}$) dt. This is the probability that the bacterium dies within an infinitesimal window of time around 5 hours, where dt is the duration of this window. For example, the probability that it lives longer than 5 hours, but shorter than (5 hours + 1 nanosecond), is (2 $hour^{-1}$)×(1 nanosecond) ≈ 6×10^{-13} (using the unit conversion 3.6×10^{12} nanoseconds = 1 hour).

There is a probability density function f with f(5 hours) = 2 $hour^{-1}$. The integral of f over any window of time (not only infinitesimal windows but also large windows) is the probability that the bacterium dies in that window.

Absolutely Continuous Univariate Distributions

A probability density function is most commonly associated with absolutely continuous univariate distributions. A random variable X has density f_X, where f_X is a non-negative Lebesgue-integrable function, if:

$$\Pr[a \le X \le b] = \int_a^b f_X(x)\,dx.$$

Hence, if F_X is the cumulative distribution function of X, then:

$$F_X(x) = \int_{-\infty}^{x} f_X(u)\,du,$$

and (if f_X is continuous at x).

Intuitively, one can think of $f_X(x)dx$ as being the probability of X falling within the infinitesimal interval $[x, x+dx]$.

Link between Discrete and Continuous Distributions

It is possible to represent certain discrete random variables as well as random variables involving both a continuous and a discrete part with a generalized probability density function, by using the Dirac delta function. (This is not possible with a probability density function in the sense defined above, it may be done with a distribution.) For example, consider a binary discrete random variable having the Rademacher distribution—that is, taking −1 or 1 for values, with probability ½ each. The density of probability associated with this variable is:

$$f(t)=\frac{1}{2}(\delta(t+1)+\delta(t-1)).$$

More generally, if a discrete variable can take *n* different values among real numbers, then the associated probability density function is:

$$f(t)=\sum_{i=1}^{n}p_i\,\delta(t-x_i),$$

where $x_1,\ldots,x_n$ are the discrete values accessible to the variable and $p_1,\ldots,p_n$ are the probabilities associated with these values.

This substantially unifies the treatment of discrete and continuous probability distributions. For instance, the above expression allows for determining statistical characteristics of such a discrete variable (such as its mean, its variance and its kurtosis), starting from the formulas given for a continuous distribution of the probability.

Families of Densities

It is common for probability density functions (and probability mass functions) to be parametrized—that is, to be characterized by unspecified parameters. For example, the normal distribution is parametrized in terms of the mean and the variance, denoted by μ and σ^2 respectively, giving the family of densities,

$$f(x;\mu,\sigma^2)=\frac{1}{\sigma\sqrt{2\pi}}e^{-\frac{1}{2}\left(\frac{x-\mu}{\sigma}\right)^2}.$$

It is important to keep in mind the difference between the domain of a family of densities and the parameters of the family. Different values of the parameters describe different distributions of different random variables on the same sample space (the same set of all possible values of the variable); this sample space is the domain of the family of random variables that this family of distributions describes. A given set of parameters describes a single distribution within the family sharing the functional form of the density. From the perspective of a given distribution, the parameters are constants, and terms in a density function that contain only parameters, but not variables, are part of the normalization factor of a distribution (the multiplicative factor that ensures that the area under the density—the probability of something in the domain occurring—equals 1). This normalization factor is outside the kernel of the distribution.

Since the parameters are constants, reparametrizing a density in terms of different parameters, to give a characterization of a different random variable in the family, means simply substituting the new parameter values into the formula in place of the old ones. Changing the domain of a probability density, however, is trickier and requires more work.

Densities Associated with Multiple Variables

For continuous random variables $X_1, \ldots, X_n$, it is also possible to define a probability density function associated to the set as a whole, often called joint probability density function. This density function is defined as a function of the n variables, such that, for any domain D in the n-dimensional space of the values of the variables $X_1, \ldots, X_n$, the probability that a realisation of the set variables falls inside the domain D is:

$$\Pr(X_1,\ldots,X_n \in D) = \int_D f_{X_1,\ldots,X_n}(x_1,\ldots,x_n)dx_1\cdots dx_n.$$

If $F(x_1, \ldots, x_n) = \Pr(X_1 \le x_1, \ldots, X_n \le x_n)$ is the cumulative distribution function of the vector $(X_1, \ldots, X_n)$, then the joint probability density function can be computed as a partial derivative:

$$f(x) = \frac{\partial^n F}{\partial x_1 \cdots \partial x_n}\bigg|_x$$

Marginal Densities

For i = 1, 2, ..., n, let $f_{Xi}(x_i)$ be the probability density function associated with variable X_i alone. This is called the marginal density function, and can be deduced from the probability density associated with the random variables $X_1, \ldots, X_n$ by integrating over all values of the other n − 1 variables:

$$f_{X_i}(x_i) = \int f(x_1,\ldots,x_n)dx_1\cdots dx_{i-1}dx_{i+1}\cdots dx_n.$$

Independence

Continuous random variables $X_1, \ldots, X_n$ admitting a joint density are all independent from each other if and only if,

$$f_{X_1,\ldots,X_n}(x_1,\ldots,x_n) = f_{X_1}(x_1)\cdots f_{X_n}(x_n).$$

Corollary

If the joint probability density function of a vector of *n* random variables can be factored into a product of *n* functions of one variable:

$$f_{X_1,\ldots,X_n}(x_1,\ldots,x_n) = f_1(x_1)\cdots f_n(x_n),$$

(where each *fi* is not necessarily a density) then the *n* variables in the set are all independent from each other, and the marginal probability density function of each of them is given by,

$$f_{X_i}(x_i) = \frac{f_i(x_i)}{\int f_i(x)dx}.$$

Example: This elementary example illustrates the above definition of multidimensional probability density functions in the simple case of a function of a set of two variables. Let us call $\vec{R}$ a 2-dimensional random vector of coordinates (*X*, *Y*): the probability to obtain $\vec{R}$ in the quarter plane of positive *x* and *y* is,

$$\Pr(X > 0, Y > 0) = \int_0^\infty \int_0^\infty f_{X,Y}(x,y)dxdy.$$

Function of Random Variables and Change of Variables in the Probability Density Function

If the probability density function of a random variable (or vector) X is given as $f_X(x)$, it is possible to calculate the probability density function of some variable Y = g(X). This is also called a "change of variable" and is in practice used to generate a random variable of arbitrary shape $f_{g(X)} = f_Y$ using a known (for instance, uniform) random number generator.

It is tempting to think that in order to find the expected value E(g(X)), one must first find the probability density $f_{g(X)}$ of the new random variable Y = g(X). However, rather than computing,

$$E(g(X)) = \int_{-\infty}^{\infty} y f_{g(X)}(y)dy,$$

one may find instead,

$$E(g(X)) = \int_{-\infty}^{\infty} g(x) f_X(x)dx.$$

The values of the two integrals are the same in all cases in which both *X* and *g*(*X*) actually have probability density functions. It is not necessary that *g* be a one-to-one function. In some cases the latter integral is computed much more easily than the former.

Scalar to Scalar

Let $g: \mathbb{R} \to \mathbb{R}$ be a monotonic function, then the resulting density function is,

$$f_Y(y) = f_X\left(g^{-1}(y)\right)\left|\frac{d}{dy}\left(g^{-1}(y)\right)\right|.$$

Here g^{-1} denotes the inverse function. This follows from the fact that the probability contained in a differential area must be invariant under change of variables. That is,

$$|f_Y(y)dy| = |f_X(x)dx|,$$

or

$$f_Y(y) = \left|\frac{dx}{dy}\right| f_X(x) = \left|\frac{d}{dy}(x)\right| f_X(x) = \left|\frac{d}{dy}\left(g^{-1}(y)\right)\right| f_X\left(g^{-1}(y)\right) = \left|\left(g^{-1}(y)\right)\right| \cdot f_X\left(g^{-1}(y)\right).$$

For functions that are not monotonic, the probability density function for y is,

$$\sum_{k=1}^{n(y)} \left|\frac{d}{dy} g_k^{-1}(y)\right| \cdot f_X\left(g_k^{-1}(y)\right),$$

where n(y) is the number of solutions in x for the equation $g(x) = y$, and $g_k^{-1}(y)$ are these solutions.

Vector to Vector

The above formulas can be generalized to variables (which we will again call y) depending on more than one other variable. $f(x_1, \ldots, x_n)$ shall denote the probability density function of the variables that y depends on, and the dependence shall be $y = g(x_1, \ldots, x_n)$. Then, the resulting density function is,

$$\int_{y=g(x_1,\ldots,x_n)} \frac{f(x_1,\ldots,x_n)}{\sqrt{\sum_{j=1}^{n} \frac{\partial g}{\partial x_j}(x_1,\ldots,x_n)^2}} dV,$$

where the integral is over the entire (n – 1)-dimensional solution of the subscripted equation and the symbolic dV must be replaced by a parametrization of this solution for a particular calculation; the variables $x_1, \ldots, x_n$ are then of course functions of this parametrization.

This derives from the following, perhaps more intuitive representation: Suppose $\mathbf{x}$ is an n-dimensional random variable with joint density f. If $\mathbf{y} = H(\mathbf{x})$, where H is a bijective, differentiable function, then $\mathbf{y}$ has density g:

$$g(\mathbf{y}) = f\left(H^{-1}(\mathbf{y})\right)\left|\det\left[\frac{dH^{-1}(\mathbf{z})}{d\mathbf{z}}\bigg|_{\mathbf{z}=\mathbf{y}}\right]\right|$$

with the differential regarded as the Jacobian of the inverse of *H(.)*, evaluated at y.

For example, in the 2-dimensional case $\mathbf{x} = (x_1, x_2)$, suppose the transform H is given as $y_1 = H_1(x_1, x_2)$, $y_2 = H_2(x_1, x_2)$ with inverses $x_1 = H_1^{-1}(y_1, y_2)$, $x_2 = H_2^{-1}(y_1, y_2)$. The joint distribution for $\mathbf{y} = (y_1, y_2)$ has density:

$$g(y_1,y_2) = f_{X_1,X_2}\left(H_1^{-1}(y_1,y_2), H_2^{-1}(y_1,y_2)\right)\left|\frac{\partial H_1^{-1}}{\partial y_1}\frac{\partial H_2^{-1}}{\partial y_2} - \frac{\partial H_1^{-1}}{\partial y_2}\frac{\partial H_2^{-1}}{\partial y_1}\right|.$$

Vector to Scalar

Let $V:\mathbb{R}^n \to \mathbb{R}$ be a differentiable function and X be a random vector taking values in $\mathbb{R}^n$, $f_X(\cdot)$ be the probability density function of X and $\delta(\cdot)$ be the Dirac delta function. It is possible to use the formulas above to determine $f_Y(\cdot)$, the probability density function of $Y = V(X)$, which will be given by,

$$f_Y(y) = \int_{\mathbb{R}^n} f_X(\mathbf{x})\delta(y - V(\mathbf{x}))d\mathbf{x}.$$

This result leads to the Law of the unconscious statistician:

$$E_Y[Y] = \int_{\mathbb{R}} y f_Y(y)dy = \int_{\mathbb{R}} y \int_{\mathbb{R}^n} f_X(\mathbf{x})\delta(y - V(\mathbf{x}))d\mathbf{x}dy =$$

$$\int_{\mathbb{R}^n}\int_{\mathbb{R}} y f_X(\mathbf{x})\delta(y - V(\mathbf{x}))dyd\mathbf{x} = \int_{\mathbb{R}^n} V(x)f_X(\mathbf{x})d\mathbf{x} = E_X[V(X)].$$

Proof:

Let Z be a collapsed random variable with probability density function $p_Z(z) = \delta(z)$ (*i.e.* a constant equal to zero). Let the random vector $\tilde{X}$ and the transform H be defined as

$$H(Z,X) = \begin{bmatrix} Z + V(X) \\ X \end{bmatrix} = \begin{bmatrix} Y \\ \tilde{X} \end{bmatrix}.$$

It is clear that H is a bijective mapping, and the Jacobian of H^{-1} is given by:

$$\frac{dH^{-1}(y,\tilde{\mathbf{x}})}{dy\,d\tilde{\mathbf{x}}} = \begin{bmatrix} 1 & -\dfrac{dV(\tilde{\mathbf{x}})}{d\tilde{\mathbf{x}}} \\ \mathbf{0}_{n\times 1} & \mathbf{I}_{n\times n} \end{bmatrix},$$

which is an upper triangular matrix with ones on the main diagonal, therefore its determinant is 1. Applying the change of variable theorem we obtain that,

$$f_{Y,X}(y,x) = f_X(\mathbf{x})\delta(y - V(\mathbf{x})),$$

which if marginalized over x leads to the desired probability density function.

Sums of Independent Random Variables

The probability density function of the sum of two independent random variables *U* and *V*, each of which has a probability density function, is the convolution of their separate density functions:

$$f_{U+V}(x) = \int_{-\infty}^{\infty} f_U(y) f_V(x-y)\,dy = (f_U * f_V)(x)$$

It is possible to generalize the previous relation to a sum of N independent random variables, with densities $U_1, \ldots, U_N$:

$$f_{U_1+\cdots+U_N}(x) = \left(f_{U_1} * \cdots * f_{U_N}\right)(x)$$

This can be derived from a two-way change of variables involving Y=U+V and Z=V, similarly to the example below for the quotient of independent random variables.

Products and Quotients of Independent Random Variables

Given two independent random variables U and V, each of which has a probability density function, the density of the product Y = UV and quotient Y=U/V can be computed by a change of variables.

Example: Quotient distribution- To compute the quotient Y = U/V of two independent random variables U and V, define the following transformation:

$$Y = U/V$$

$$Z = V$$

Then, the joint density p(y,z) can be computed by a change of variables from U,V to Y,Z, and Y can be derived by marginalizing out Z from the joint density.

The inverse transformation is,

$$U = YZ$$

$$V = Z$$

The Jacobian matrix $J(U,V \mid Y,Z)$ of this transformation is

$$\begin{vmatrix} \frac{\partial u}{\partial y} & \frac{\partial u}{\partial z} \\ \frac{\partial v}{\partial y} & \frac{\partial v}{\partial z} \end{vmatrix} = \begin{vmatrix} z & y \\ 0 & 1 \end{vmatrix} = |z|.$$

Thus:

$$p(y,z) = p(u,v)\ J(u,v \mid y,z) = p(u)p(v)\ J(u,v \mid y,z) = p_U(yz)p_V(z)|z|.$$

And the distribution of *Y* can be computed by marginalizing out *Z*:

$$p(y) = \int_{-\infty}^{\infty} p_U(yz)p_V(z)|z|dz$$

This method crucially requires that the transformation from U,V to Y,Z be bijective. The above transformation meets this because Z can be mapped directly back to V, and for a given V the quotient U/V is monotonic. This is similarly the case for the sum U + V, difference U – V and product UV.

Exactly the same method can be used to compute the distribution of other functions of multiple independent random variables.

Example: Quotient of two standard normals- Given two standard normal variables U and V, the quotient can be computed as follows. First, the variables have the following density functions:

$$p(u) = \frac{1}{\sqrt{2\pi}} e^{-\frac{u^2}{2}}$$

$$p(v) = \frac{1}{\sqrt{2\pi}} e^{-\frac{v^2}{2}}$$

We transform as described above:

$$Y = U/V$$

$$Z = V$$

This leads to:

$$\begin{aligned}
p(y) &= \int_{-\infty}^{\infty} p_U(yz)p_V(z)|z|dz \\
&= \int_{-\infty}^{\infty} \frac{1}{\sqrt{2\pi}} e^{-\frac{1}{2}y^2z^2} \frac{1}{\sqrt{2\pi}} e^{-\frac{1}{2}z^2} |z|dz \\
&= \int_{-\infty}^{\infty} \frac{1}{2\pi} e^{-\frac{1}{2}(y^2+1)z^2} |z|dz \qquad = 2\int_{0}^{\infty} \frac{1}{2\pi} e^{-\frac{1}{2}(y^2+1)z^2} zdz \\
&= \int_{0}^{\infty} \frac{1}{\pi} e^{-(y^2+1)u} du \qquad u = \tfrac{1}{2}z^2 \\
&= -\frac{1}{\pi(y^2+1)} e^{-(y^2+1)u} \Big]_{u=0}^{\infty} \qquad = \frac{1}{\pi(y^2+1)}
\end{aligned}$$

This is the density of a standard Cauchy distribution.

Discrete Distributions

A statistical distribution whose variables can take on only discrete values. Abramowitz and Stegun give a table of the parameters of most common discrete distributions.

A discrete distribution with probability function $P(x_k)$ defined over k = 1, 2, ..., N has distribution function,

$$D(x_n) = \sum_{(k=1)}^{n} P(x_k)$$

and population mean,

$$\mu = \frac{1}{N}\sum_{(k=1)}^{N} x_k P(x_k).$$

Discrete Distributions with Finite Support

Bernoulli Distribution

In probability theory and statistics, the Bernoulli distribution, named after Swiss mathematician Jacob Bernoulli, is the discrete probability distribution of a random variable which takes the value 1 with probability p and the value 0 with probability $q = 1 - p$, that is, the probability distribution of any single experiment that asks a yes–no question; the question results in a boolean-valued outcome, a single bit whose value is success/yes/true/one with probability p and failure/no/false/zero with probability q. It can be used to represent a (possibly biased) coin toss where 1 and 0 would represent "heads" and "tails" (or vice versa), respectively, and p would be the probability of the coin landing on heads or tails, respectively. In particular, unfair coins would have $p \neq 1/2$.

The Bernoulli distribution is a special case of the binomial distribution where a single trial is conducted (so *n* would be 1 for such a binomial distribution). It is also a special case of the two-point distribution, for which the possible outcomes need not be 0 and 1.

Bernoulli	
Parameters	$0 \le p \le 1$ $q = 1 - p$
Support	$k \in \{0,1\}$
pmf	$\begin{cases} q = 1 - p & \text{if } k = 0 \\ p & \text{if } k = 1 \end{cases}$

CDF	$\begin{cases} 0 & \text{if } k < 0 \\ 1-p & \text{if } 0 \le k < 1 \\ 1 & \text{if } k \ge 1 \end{cases}$
Mean	p
Median	$\begin{cases} 0 & \text{if } p < 1/2 \\ [0,1] & \text{if } p = 1/2 \\ 1 & \text{if } p > 1/2 \end{cases}$
Mode	$\begin{cases} 0 & \text{if } p < 1/2 \\ 0,1 & \text{if } p = 1/2 \\ 1 & \text{if } p > 1/2 \end{cases}$
Variance	$p(1-p) = pq$
Skewness	$\frac{q-p}{\sqrt{pq}}$
Ex. kurtosis	$\frac{1-6pq}{pq}$
Entropy	$-q\ln q - p\ln p$
MGF	$q + pe^{t}$
CF	$q + pe^{it}$
PGF	$q + pz$
Fisher information	$\frac{1}{pq}$

Properties

If X is a random variable with this distribution, then:

$$\Pr(X=1) = p = 1 - \Pr(X=0) = 1 - q.$$

The probability mass function f of this distribution, over possible outcomes *k*, is,

$$f(k;p) = \begin{cases} p & \text{if } k = 1, \\ q = 1-p & \text{if } k = 0. \end{cases}$$

This can also be expressed as,

$$f(k;p) = p^k(1-p)^{1-k} \quad \text{for } k \in \{0,1\}$$

or as,

$$f(k;p) = pk + (1-p)(1-k) \quad \text{for } k \in \{0,1\}.$$

The Bernoulli distribution is a special case of the binomial distribution with $n = 1$. The kurtosis goes to infinity for high and low values of p, but for $p = 1/2$ the two-point distributions including the Bernoulli distribution have a lower excess kurtosis than any other probability distribution, namely -2.

The Bernoulli distributions for $0 \le p \le 1$ form an exponential family. The maximum likelihood estimator of p based on a random sample is the sample mean.

Mean

The expected value of a Bernoulli random variable X is,

$$E(X) = p$$

This is due to the fact that for a Bernoulli distributed random variable X with $\Pr(X = 1) = p$ and $\Pr(X = 0) = q$ we find,

$$E[X] = \Pr(X = 1)\cdot 1 + \Pr(X = 0)\cdot 0 = p\cdot 1 + q\cdot 0 = p.$$

Variance

The variance of a Bernoulli distributed X is,

$$\mathrm{Var}[X] = pq = p(1-p)$$

We first find,

$$E[X^2] = \Pr(X = 1)\cdot 1^2 + \Pr(X = 0)\cdot 0^2 = p\cdot 1^2 + q\cdot 0^2 = p$$

From this follows,

$$\mathrm{Var}[X] = E[X^2] - E[X]^2 = p - p^2 = p(1-p) = pq$$

Skewness

The skewness is $\frac{q-p}{\sqrt{pq}} = \frac{1-2p}{\sqrt{pq}}$. When we take the standardized Bernoulli distributed random variable $\frac{X-E[X]}{\sqrt{Var[X]}}$ we find that this random variable attains $\frac{q}{\sqrt{pq}}$ with probability p and attains $-\frac{p}{\sqrt{pq}}$ with probability q. Thus we get

$$\begin{aligned}\gamma_1 &= E\left[\left(\frac{X-E[X]}{\sqrt{Var[X]}}\right)^3\right]\\ &= p\cdot\left(\frac{q}{\sqrt{pq}}\right)^3 + q\cdot\left(-\frac{p}{\sqrt{pq}}\right)^3\\ &= \frac{1}{\sqrt{pq}^3}\left(pq^3 - qp^3\right)\\ &= \frac{pq}{\sqrt{pq}^3}(q-p)\\ &= \frac{q-p}{\sqrt{pq}}\end{aligned}$$

Higher Moments and Cumulants

The central moment of order k is given by,

$$\mu_k = (1-p)(-p)^k + p(1-p)^k.$$

The first six central moments are,

$$\begin{aligned}\mu_1 &= 0,\\ \mu_2 &= p(1-p),\\ \mu_3 &= p(1-p)(1-2p),\\ \mu_4 &= p(1-p)(1-3p(1-p)),\\ \mu_5 &= p(1-p)(1-2p)(1-2p(1-p)),\\ \mu_6 &= p(1-p)(1-5p(1-p)(1-p(1-p))).\end{aligned}$$

The higher central moments can be expressed more compactly in terms of μ_2 and μ_3,

$$\begin{aligned}\mu_4 &= \mu_2(1-3\mu_2),\\ \mu_5 &= \mu_3(1-2\mu_2),\\ \mu_6 &= \mu_2(1-5\mu_2(1-\mu_2)).\end{aligned}$$

The first six cumulants are,

$$\kappa_1 = 0,$$
$$\kappa_2 = \mu_2,$$
$$\kappa_3 = \mu_3,$$
$$\kappa_4 = \mu_2(1-6\mu_2),$$
$$\kappa_5 = \mu_3(1-12\mu_2),$$
$$\kappa_6 = \mu_2(1-30\mu_2(1-4\mu_2)).$$

Related Distributions

- If $X_1,\ldots,X_n$ are independent, identically distributed (i.i.d.) random variables, all Bernoulli trials with success probability *p*, then their sum is distributed according to a binomial distribution with parameters *n* and *p*:
  $\sum_{k=1}^{n} X_k \sim B(n,p)$ (binomial distribution).
  The Bernoulli distribution is simply $B(1,p)$, also written as $\text{Bernoulli}(p)$.
- The categorical distribution is the generalization of the Bernoulli distribution for variables with any constant number of discrete values.
- The Beta distribution is the conjugate prior of the Bernoulli distribution.
- The geometric distribution models the number of independent and identical Bernoulli trials needed to get one success.
- If $Y \sim \text{Bernoulli}\left(\frac{1}{2}\right)$, then $2Y-1$ has a Rademacher distribution.

Binomial Distribution

In probability theory and statistics, the binomial distribution with parameters n and p is the discrete probability distribution of the number of successes in a sequence of n independent experiments, each asking a yes–no question, and each with its own boolean-valued outcome: success/yes/true/one (with probability p) or failure/no/false/zero (with probability q = 1 − p). A single success/failure experiment is also called a Bernoulli trial or Bernoulli experiment and a sequence of outcomes is called a Bernoulli process; for a single trial, i.e., n = 1, the binomial distribution is a Bernoulli distribution. The binomial distribution is the basis for the popular binomial test of statistical significance.

The binomial distribution is frequently used to model the number of successes in a sample of size *n* drawn with replacement from a population of size N. If the sampling is carried out without replacement, the draws are not independent and so the resulting distribution is a hypergeometric distribution, not a binomial one. However, for N much larger than n, the binomial distribution remains a good approximation, and is widely used.

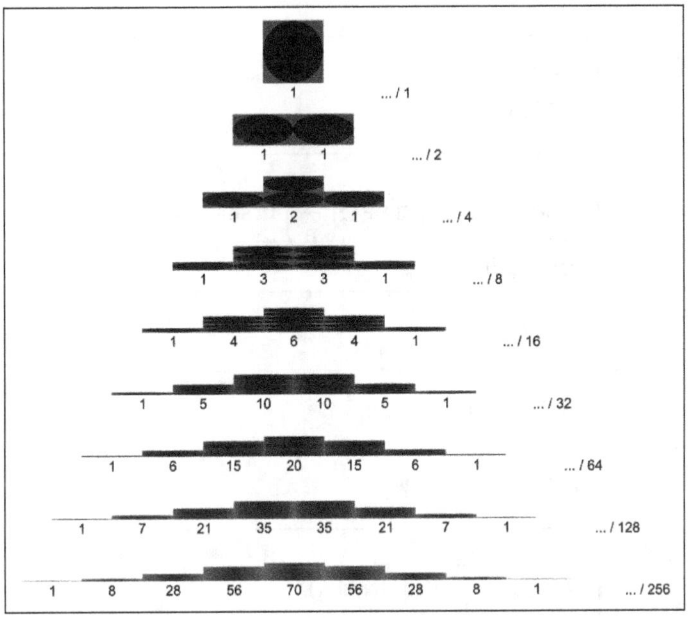

Binomial distribution for $p=0.5$ with n and k as in Pascal›s triangle. The probability that a ball in a Galton box with 8 layers (n = 8) ends up in the central bin (k = 4) is $70/256$.

Binomial Distribution	
Notation	$B(n,p)$
Parameters	$n \in \{0,1,2,\ldots\}$ – number of trials $p \in [0,1]$ – success probability for each trial $q = 1-p$
Support	$k \in \{0,1,\ldots,n\}$ – number of successes
pmf	$\binom{n}{k} p^k q^{n-k}$
CDF	$I_q(n-k,1+k)$
Mean	np
Median	$\lfloor np \rfloor$ or $\lceil np \rceil$
Mode	$\lfloor (n+1)p \rfloor$ or $\lceil (n+1)p \rceil - 1$
Variance	npq

Skewness	$\frac{q-p}{\sqrt{npq}}$
Ex. kurtosis	$\frac{1-6pq}{npq}$
Entropy	$\frac{1}{2}\log_2(2\pi enpq)+O\left(\frac{1}{n}\right)$ in shannons. For nats, use the natural log in the log.
MGF	$(q+pe^t)^n$
CF	$(q+pe^{it})^n$
PGF	$G(z)=[q+pz]^n$
Fisher information	$g_n(p)=\frac{n}{pq}$ (for fixed n).

In general, if the random variable X follows the binomial distribution with parameters $n \in N$ and $p \in [0,1]$, we write $X \sim B(n, p)$. The probability of getting exactly *k* successes in *n* independent Bernoulli trials is given by the probability mass function:

$$f(k,n,p)=\Pr(k;n,p)=\Pr(X=k)=\binom{n}{k}p^k(1-p)^{n-k}$$

for *k* = 0, 1, 2, ..., n, where

$$\binom{n}{k}=\frac{n!}{k!(n-k)!}$$

is the binomial coefficient, hence the name of the distribution. The formula can be understood as follows. *k* successes occur with probability p^k and $n - k$ failures occur with probability $(1 - p)^{n-k}$. However, the k successes can occur anywhere among the *n* trials, and there are $\binom{n}{k}$ different ways of distributing *k* successes in a sequence of *n* trials.

In creating reference tables for binomial distribution probability, usually the table is filled in up to n/2 values. This is because for k > n/2, the probability can be calculated by its complement as,

$$f(k,n,p)=f(n-k,n,1-p).$$

Looking at the expression f(k, n, p) as a function of *k*, there is a *k* value that maximizes it. This *k* value can be found by calculating,

$$\frac{f(k+1,n,p)}{f(k,n,p)} = \frac{(n-k)p}{(k+1)(1-p)}$$

and comparing it to 1. There is always an integer *M* that satisfies,

$$(n+1)p-1 \le M < (n+1)p.$$

f(k, n, p) is monotone increasing for k < M and monotone decreasing for k > M, with the exception of the case where (n + 1)p is an integer. In this case, there are two values for which f is maximal: (n + 1)p and (n + 1)p − 1. M is the most probable outcome (that is, the most likely, although this can still be unlikely overall) of the Bernoulli trials and is called the mode.

Cumulative Distribution Function

The cumulative distribution function can be expressed as:

$$F(k;n,p) = \Pr(X \le k) = \sum_{i=0}^{\lfloor k \rfloor} \binom{n}{i} p^i (1-p)^{n-i}$$

where $\lfloor k \rfloor$ is the "floor" under *k*, i.e. the greatest integer less than or equal to *k*.

It can also be represented in terms of the regularized incomplete beta function, as follows:

$$\begin{aligned} F(k;n,p) &= \Pr(X \le k) \\ &= I_{1-p}(n-k,k+1) \\ &= (n-k)\binom{n}{k}\int_0^{1-p} t^{n-k-1}(1-t)^k\,dt. \end{aligned}$$

which is equivalent to the cumulative distribution function of the F-distribution

$$F(k;n,p) = F_{F\text{-distribution}}\left(x = \frac{1-p}{p}\frac{k+1}{n-k}; d_1 = 2(n-k), d_2 = 2(k+1)\right).$$

Some closed-form bounds for the cumulative distribution function are given below:

Examples: Suppose a biased coin comes up heads with probability 0.3 when tossed. What is the probability of achieving 0, 1,..., 6 heads after six tosses?

$$\Pr(0 \text{ heads}) = f(0) = \Pr(X=0) = \binom{6}{0} 0.3^0 (1-0.3)^{6-0} = 0.117649$$

$$\Pr(1\text{ heads}) = f(1) = \Pr(X=1) = \binom{6}{1} 0.3^1 (1-0.3)^{6-1} = 0.302526$$

$$\Pr(2\text{ heads}) = f(2) = \Pr(X=2) = \binom{6}{2} 0.3^2 (1-0.3)^{6-2} = 0.324135$$

$$\Pr(3\text{ heads}) = f(3) = \Pr(X=3) = \binom{6}{3} 0.3^3 (1-0.3)^{6-3} = 0.18522$$

$$\Pr(4\text{ heads}) = f(4) = \Pr(X=4) = \binom{6}{4} 0.3^4 (1-0.3)^{6-4} = 0.059535$$

$$\Pr(5\text{ heads}) = f(5) = \Pr(X=5) = \binom{6}{5} 0.3^5 (1-0.3)^{6-5} = 0.010206$$

$$\Pr(6\text{ heads}) = f(6) = \Pr(X=6) = \binom{6}{6} 0.3^6 (1-0.3)^{6-6} = 0.000729$$

Properties

Expected Value and Variance

If $X \sim B(n, p)$, that is, X is a binomially distributed random variable, n being the total number of experiments and p the probability of each experiment yielding a successful result, then the expected value of X is:

$$E[X] = np.$$

This follows from the linearity of the expected value along with fact that X is the sum of n identical Bernoulli random variables, each with expected value p. In other words, if $X_1,\ldots,X_n$ are identical (and independent) Bernoulli random variables with parameter p, then $X = X_1 + \cdots + X_n$ and

$$E[X] = E[X_1 + \cdots + X_n] = E[X_1] + \cdots + E[X_n] = p + \cdots + p = np.$$

The variance is:

$$\mathrm{Var}(X) = np(1-p).$$

This similarly follows from the fact that the variance of a sum of independent random variables is the sum of the variances.

Higher Moments

The first 6 central moments are given by,

$$
\begin{aligned}
\mu_1 &= 0,\\
\mu_2 &= np(1-p),\\
\mu_3 &= np(1-p)(1-2p),\\
\mu_4 &= np(1-p)(1+(3n-6)p(1-p)),\\
\mu_5 &= np(1-p)(1-2p)(1+(10n-12)p(1-p)),\\
\mu_6 &= np(1-p)(1-30p(1-p)(1-4p(1-p))+5np(1-p)(5-26p(1-p))+15n^2p^2(1-p)^2).
\end{aligned}
$$

Mode

Usually the mode of a binomial $B(n,p)$ distribution is equal to $\lfloor (n+1)p \rfloor$, where $\lfloor \cdot \rfloor$ is the floor function. However, when (n + 1)p is an integer and p is neither 0 nor 1, then the distribution has two modes: (n + 1)p and (n + 1)p – 1. When p is equal to 0 or 1, the mode will be 0 and n correspondingly. These cases can be summarized as follows:

$$
\text{mode} = \begin{cases} \lfloor (n+1)p \rfloor & \text{if } (n+1)p \text{ is 0 or a noninteger},\\ (n+1)p \ \text{ and } \ (n+1)p-1 & \text{if } (n+1)p \in \{1,\ldots,n\},\\ n & \text{if } (n+1)p = n+1. \end{cases}
$$

Proof: Let,

$$
f(k) = \binom{n}{k} p^k q^{n-k}.
$$

For $p=0$ only $f(0)$ has a nonzero value with $f(0)=1$. For $p=1$ we find $f(n)=1$ and $f(k)=0$ for $k \neq n$. This proves that the mode is 0 for $p=0$ and n for $p=1$.

Let $0<p<1$. We find,

$$
\frac{f(k+1)}{f(k)} = \frac{(n-k)p}{(k+1)(1-p)}.
$$

From this follows:

$$
\begin{aligned}
k > (n+1)p-1 &\Rightarrow f(k+1) < f(k)\\
k = (n+1)p-1 &\Rightarrow f(k+1) = f(k)\\
k < (n+1)p-1 &\Rightarrow f(k+1) > f(k)
\end{aligned}
$$

So when $(n+1)p-1$ is an integer, then $(n+1)p-1$ and $(n+1)p$ is a mode. In the case that $(n+1)p-1 \notin \mathbb{Z}$, then only $\lfloor (n+1)p-1 \rfloor + 1 = \lfloor (n+1)p \rfloor$ is a mode.

Median

In general, there is no single formula to find the median for a binomial distribution, and it may even be non-unique. However several special results have been established:

- If np is an integer, then the mean, median, and mode coincide and equal np.
- Any median m must lie within the interval $\lfloor np \rfloor \leq m \leq \lceil np \rceil$.
- A median m cannot lie too far away from the mean: $|m - np| \leq \min\{\ln 2, \max\{p, 1 - p\}\}$.
- The median is unique and equal to m = round(np) when $|m - np| \leq \min\{p, 1 - p\}$ (except for the case when $p=\frac{1}{2}$ and n is odd).
- When $p = 1/2$ and n is odd, any number m in the interval $\frac{1}{2}(n-1) \leq m \leq \frac{1}{2}(n+1)$ is a median of the binomial distribution. If $p = 1/2$ and n is even, then $m = n/2$ is the unique median.

Tail Bounds

For $k \leq np$, upper bounds for the lower tail of the distribution function can be derived. Recall that $F(k;n,p) = \Pr(X \leq k)$, the probability that there are at most k successes.

Hoeffding's inequality yields the bound,

$$F(k;n,p) \leq \exp\left(-2\frac{(np-k)^2}{n}\right),$$

and Chernoff's inequality can be used to derive the bound,

$$F(k;n,p) \leq \exp\left(-\frac{1}{2p}\frac{(np-k)^2}{n}\right).$$

Moreover, these bounds are reasonably tight when $p = 1/2$, since the following expression holds for all $k \geq 3n/8$,

$$F\left(k;n,\tfrac{1}{2}\right) \leq \frac{14}{15}\exp\left(-\frac{16\left(\frac{n}{2}-k\right)^2}{n}\right).$$

However, the bounds do not work well for extreme values of p. In particular, as $p \to 1$, value F(k;n,p) goes to zero (for fixed k, n with k < n) while the upper bound above goes to a positive constant. In this case a better bound is given by,

$$F(k;n,p) \leq \exp\left(-nD\left(\frac{k}{n} \| p\right)\right) \quad \text{if } 0 < \frac{k}{n} < p$$

where D(a || p) is the relative entropy between an a-coin and a p-coin (i.e. between the Bernoulli(a) and Bernoulli(p) distribution):

$$D(a \| p) = (a)\log\frac{a}{p} + (1-a)\log\frac{1-a}{1-p}.$$

Asymptotically, this bound is reasonably tight. An equivalent formulation of the bound is,

$$\Pr(X \geq k) = F(n-k; n, 1-p) \leq \exp\left(-nD\left(\frac{k}{n} \| p\right)\right) \quad \text{if } p < \frac{k}{n} < 1.$$

Both these bounds are derived directly from the Chernoff bound. It can also be shown that,

$$\Pr(X \geq k) = F(n-k; n, 1-p) \geq \frac{1}{(n+1)^2}\exp\left(-nD\left(\frac{k}{n} \| p\right)\right) \quad \text{if } p < \frac{k}{n} < 1.$$

This is proved using the method of types.

We can also change the $(n+1)^2$ in the denominator to $\sqrt{2n}$, by approximating the binomial coefficient with Stirling's formula.

Additionally, the $(n+1)^2$ in the denominator can also be changed to,

$$\max\left(2, \sqrt{4\pi n D\left(\frac{k}{n} \| p\right)}\right)$$

by exploiting relationships between the binomial and Standard Normal Distributions, and applying an approximation of Mills' ratio.

Statistical Inference

Estimation of Parameters

When n is known, the parameter p can be estimated using the proportion of successes: This estimator is found using maximum likelihood estimator and also the method of moments. This estimator is unbiased and uniformly with minimum variance, proven using Lehmann–Scheffé theorem, since it is based on a minimal sufficient and complete statistic (i.e.: x). It is also consistent both in probability and in MSE.

A closed form Bayes estimator for *p* also exists when using the Beta distribution as a conjugate prior distribution. When using a general $\text{Beta}(\alpha,\beta)$ as a prior, the posterior mean estimator is: $\hat{p}_b = \frac{x+\alpha}{n+\alpha+\beta}$. The Bayes estimator is asymptoticly efficient and as the sample size approaches infinity ($n \to \infty$), it approaches the MLE solution. The Bayes estimator is biased (how much depends on the priors), admissible and consistent in probability.

For the special case of using the standard uniform distribution as a non-informative prior $\text{Beta}(\alpha=1,\beta=1)=U(0,1)$, the posterior mean estimator becomes $\hat{p}_b=\frac{x+1}{n+2}$ (a posterior mode should just lead to the standard estimator). This method is called the rule of succession, which was introduced in the 18th century by Pierre-Simon Laplace.

When estimating *p* with very rare events and a small *n* (e.g.: if x=0), then using the standard estimator leads to $\hat{p}=0$, which sometimes is unrealistic and undesirable. In such cases there are various alternative estimators. One way is to use the Bayes estimator, leading to: $\hat{p}_b=\frac{1}{n+2}$). Another method is to use the upper bound of the confidence interval obtained using the rule of three: $\hat{p}_{\text{rule of 3}}=\frac{3}{n}$).

Confidence Intervals

Even for quite large values of *n*, the actual distribution of the mean is significantly non-normal. Because of this problem several methods to estimate confidence intervals have been proposed.

In the equations for confidence intervals below, the variables have the following meaning:

- n_1 is the number of successes out of *n*, the total number of trials.
- $\hat{p}=\frac{n_1}{n}$ is the proportion of successes.
- z is the $1-\frac{1}{2}\alpha$ quantile of a standard normal distribution (i.e., probit) corresponding to the target error rate α. For example, for a 95% confidence level the error $\alpha=0.05$, so $1-\frac{1}{2}\alpha=0.975$ and $z=1.96$.

Wald method:

$$\hat{p}\pm z\sqrt{\frac{\hat{p}(1-\hat{p})}{n}}.$$

A continuity correction of 0.5/*n* may be added.

Agresti–coull method:

$$\tilde{p}\pm z\sqrt{\frac{\tilde{p}(1-\tilde{p})}{n+z^2}}.$$

Here the estimate of *p* is modified to,

$$\frac{n \qquad -z}{n \quad z}$$

Arcsine method:

$$\sin^2\left(\arcsin\left(\sqrt{\hat{p}}\right) \pm \frac{z}{2\sqrt{n}}\right).$$

Wilson (score) method:

The notation in the formula below differs from the previous formulas in two respects:

- Firstly, z_x has a slightly different interpretation in the formula below: It has its ordinary meaning of 'the xth quantile of the standard normal distribution', rather than being a shorthand for 'the (1 – x)-th quantile'.

  Secondly, this formula does not use a plus-minus to define the two bounds. Instead, one may use $z = z_{\alpha/2}$ to get the lower bound, or use $z = z_{1-\alpha/2}$ to get the upper bound. For example: For a 95% confidence level the error $z = z_{\alpha/2} = z_{0.025} = -1.96$, = 0.05, so one gets the lower bound by using $z = z_{1-\alpha/2} = z_{0.975} = 1.96$ and one gets the upper bound by using,

$$\frac{\hat{p} + \frac{z^2}{2n} + z\sqrt{\frac{\hat{p}(1-\hat{p})}{n} + \frac{z^2}{4n^2}}}{1 + \frac{z^2}{n}}.$$

Comparison

The exact (Clopper–Pearson) method is the most conservative. The Wald method, although commonly recommended in textbooks, is the most biased.

Related Distributions

Sums of Binomials

If X ~ B(n, p) and Y ~ B(m, p) are independent binomial variables with the same probability p, then X + Y is again a binomial variable; its distribution is Z=X+Y ~ B(n+m, p):

$$P(Z=k) = \sum_{i=0}^{k}\left[\binom{n}{i} p^i (1-p)^{n-i}\right]\left[\binom{m}{k-i} p^{k-i} (1-p)^{m-k+i}\right]$$

$$= \binom{n+m}{k} p^k (1-p)^{n+m-k}$$

However, if X and Y do not have the same probability p, then the variance of the sum will be smaller than the variance of a binomial variable distributed as $B(n+m,\bar{p})$.

Ratio of Two Binomial Distributions

This result was first derived by Katz et al. in 1978. Let $X \sim B(n,p_1)$ and $Y \sim B(m,p_2)$ be independent. Let $T = (X/n)/(Y/m)$. Then $\log(T)$ is approximately normally distributed with mean $\log(p_1/p_2)$ and variance $((1/p_1) - 1)/n + ((1/p_2) - 1)/m$.

Conditional Binomials

If $X \sim B(n, p)$ and $Y \mid X \sim B(X, q)$ (the conditional distribution of Y, given X), then Y is a simple binomial random variable with distribution $Y \sim B(n, pq)$.

For example, imagine throwing n balls to a basket U_X and taking the balls that hit and throwing them to another basket U_Y. If p is the probability to hit U_X then $X \sim B(n, p)$ is the number of balls that hit U_X. If q is the probability to hit U_Y then the number of balls that hit U_Y is $Y \sim B(X, q)$ and therefore $Y \sim B(n, pq)$.

Proof: Since $X \sim B(n,p)$ and $Y \sim B(X,q)$, by the law of total probability,

$$\begin{aligned}\Pr[Y=m] &= \sum_{k=m}^{n} \Pr[Y=m \mid X=k]\Pr[X=k] \\ &= \sum_{k=m}^{n} \binom{n}{k}\binom{k}{m} p^k q^m (1-p)^{n-k}(1-q)^{k-m}\end{aligned}$$

Since $\binom{n}{k}\binom{k}{m} = \binom{n}{m}\binom{n-m}{k-m}$, the equation above can be expressed as,

$$\Pr[Y=m] = \sum_{k=m}^{n} \binom{n}{m}\binom{n-m}{k-m} p^k q^m (1-p)^{n-k}(1-q)^{k-m}$$

Factoring $p^k = p^m p^{k-m}$ and pulling all the terms that don't depend on k out of the sum now yields

$$\begin{aligned}\Pr[Y=m] &= \binom{n}{m} p^m q^m \left(\sum_{k=m}^{n} \binom{n-m}{k-m} p^{k-m}(1-p)^{n-k}(1-q)^{k-m}\right) \\ &= \binom{n}{m} (pq)^m \left(\sum_{k=m}^{n} \binom{n-m}{k-m} (p(1-q))^{k-m}(1-p)^{n-k}\right)\end{aligned}$$

After substituting $i = k-m$ in the expression above, we get,

$$\Pr[Y=m] = \binom{n}{m} (pq)^m \left(\sum_{i=0}^{n-m} \binom{n-m}{i} (p-pq)^i (1-p)^{n-m-i}\right)$$

The sum (in the parentheses) above equals $p-pq+1-p)^{n-m}$ by the binomial theorem. Substituting this in finally yields,

$$\Pr[Y=m]=\binom{n}{m}(pq)^m(p-pq+1-p)^{n-m}$$

$$=\binom{n}{m}(pq)^m(1-pq)^{n-m}$$

and thus $Y \sim B(n,pq)$ as desired.

The Bernoulli distribution is a special case of the binomial distribution, where n = 1. Symbolically, X ~ B(1, p) has the same meaning as X ~ Bernoulli(p). Conversely, any binomial distribution, B(n, p), is the distribution of the sum of n Bernoulli trials, Bernoulli(p), each with the same probability p.

Poisson Binomial Distribution

The binomial distribution is a special case of the Poisson binomial distribution, or general binomial distribution, which is the distribution of a sum of n independent non-identical Bernoulli trials B(pi).

Normal Approximation

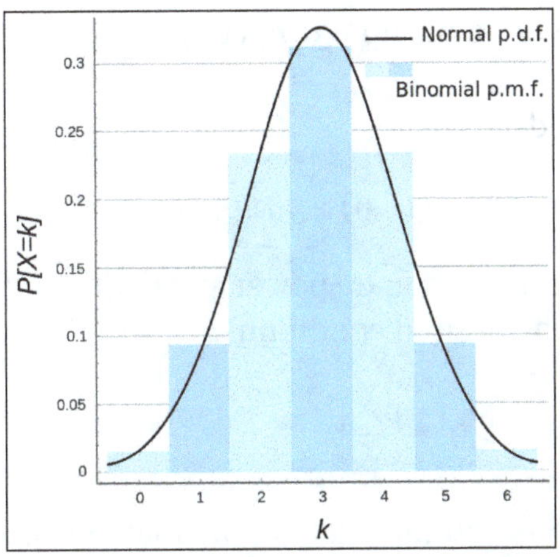

Binomial probability mass function and normal probability density function approximation for n = 6 and p = 0.5.

If n is large enough, then the skew of the distribution is not too great. In this case a reasonable approximation to B(n, p) is given by the normal distribution,

$$\mathcal{N}(np,np(1-p)),$$

and this basic approximation can be improved in a simple way by using a suitable continuity correction. The basic approximation generally improves as n increases (at least

20) and is better when p is not near to 0 or 1. Various rules of thumb may be used to decide whether n is large enough, and p is far enough from the extremes of zero or one:

- One rule is that for $n > 5$ the normal approximation is adequate if the absolute value of the skewness is strictly less than 1/3; that is, if,

$$\frac{|1-2p|}{\sqrt{np(1-p)}}=\frac{1}{\sqrt{n}}\left|\sqrt{\frac{1-p}{p}}-\sqrt{\frac{p}{1-p}}\right|<\frac{1}{3}.$$

- A stronger rule states that the normal approximation is appropriate only if everything within 3 standard deviations of its mean is within the range of possible values; that is, only if,

$$\mu\pm3\sigma=np\pm3\sqrt{np(1-p)}\in(0,n).$$

This 3-standard-deviation rule is equivalent to the following conditions, which also imply the first rule above.

$$n>9\left(\frac{1-p}{p}\right)\quad\text{and}\quad n>9\left(\frac{p}{1-p}\right).$$

Proof: The rule $np\pm3\sqrt{np(1-p)}\in(0,n)$ is totally equivalent to request that,

$$np-3\sqrt{np(1-p)}>0\quad\text{and}\quad np+3\sqrt{np(1-p)}<n.$$

Moving terms around yields:

$$np>3\sqrt{np(1-p)}\quad\text{and}\quad n(1-p)>3\sqrt{np(1-p)}.$$

Since $0<p<1$, , we can apply the square power and divide by the respective factors np^2 and $n(1-p)^2$, to obtain the desired conditions:

$$n>9\left(\frac{1-p}{p}\right)\quad\text{and}\quad n>9\left(\frac{p}{1-p}\right).$$

Notice that these conditions automatically imply that $n>9$. On the other hand, apply again the square root and divide by 3,

$$\frac{\sqrt{n}}{3}>\sqrt{\frac{1-p}{p}}>0\quad\text{and}\quad\frac{\sqrt{n}}{3}>\sqrt{\frac{p}{1-p}}>0.$$

Subtracting the second set of inequalities from the first one yields:

$$\frac{\sqrt{n}}{3}>\sqrt{\frac{1-p}{p}}-\sqrt{\frac{p}{1-p}}>-\frac{\sqrt{n}}{3};$$

and so, the desired first rule is satisfied,

$$\left|\sqrt{\frac{1-p}{p}}-\sqrt{\frac{p}{1-p}}\right|<\frac{\sqrt{n}}{3}.$$

- Another commonly used rule is that both values np and n(1−p) must be greater than or equal to 5. However, the specific number varies from source to source, and depends on how good an approximation one wants. In particular, if one uses 9 instead of 5.

Proof: Assume that both values np and $n(1-p)$ are greater than 9. Since $0<p<1$, we easily have that,

$$np \geq 9 > 9(1-p) \quad and \quad n(1-p) \geq 9 > 9p.$$

We only have to divide now by the respective factors p and $1-p$, to deduce the alternative form of the 3-standard-deviation rule:

$$n > 9\left(\frac{1-p}{p}\right) \quad and \quad n > 9\left(\frac{p}{1-p}\right).$$

The following is an example of applying a continuity correction. Suppose one wishes to calculate Pr(X ≤ 8) for a binomial random variable X. If Y has a distribution given by the normal approximation, then Pr(X ≤ 8) is approximated by Pr(Y ≤ 8.5). The addition of 0.5 is the continuity correction; the uncorrected normal approximation gives considerably less accurate results.

This approximation, known as de Moivre–Laplace theorem, is a huge time-saver when undertaking calculations by hand (exact calculations with large n are very onerous); historically, it was the first use of the normal distribution. Nowadays, it can be seen as a consequence of the central limit theorem since B(n, p) is a sum of n independent, identically distributed Bernoulli variables with parameter p. This fact is the basis of a hypothesis test, a "proportion z-test", for the value of p using x/n, the sample proportion and estimator of p, in a common test statistic.

For example, suppose one randomly samples n people out of a large population and ask them whether they agree with a certain statement. The proportion of people who agree will of course depend on the sample. If groups of n people were sampled repeatedly and truly randomly, the proportions would follow an approximate normal distribution with mean equal to the true proportion p of agreement in the population and with standard deviation $\sigma=\sqrt{\frac{p(1-p)}{n}}$.

Poisson Approximation

The binomial distribution converges towards the Poisson distribution as the number of trials goes to infinity while the product np remains fixed or at least p tends to zero.

Therefore, the Poisson distribution with parameter $\lambda = np$ can be used as an approximation to B(n, p) of the binomial distribution if n is sufficiently large and p is sufficiently small. According to two rules of thumb, this approximation is good if $n \geq 20$ and $p \leq 0.05$, or if $n \geq 100$ and $np \leq 10$.

Limiting Distributions

- Poisson limit theorem: As n approaches ∞ and p approaches 0 with the product np held fixed, the Binomial(n, p) distribution approaches the Poisson distribution with expected value $\lambda = np$.
- de Moivre–Laplace theorem: As n approaches ∞ while p remains fixed, the distribution of, $\frac{X-np}{\sqrt{np(1-p)}}$ approaches the normal distribution with expected value 0 and variance 1. This result is sometimes loosely stated by saying that the distribution of X is asymptotically normal with expected value np and variance np(1 – p). This result is a specific case of the central limit theorem.

Beta Distribution

The binomial distribution and beta distribution are different views of the same model of repeated Bernoulli trials. The binomial distribution is the PMF of k successes given n independent events each with a probability p of success. Mathematically, when $\alpha = k+1$ and $\beta = n-k+1$, the beta distribution and the binomial distribution are related by a factor of n+1:

$$\mathrm{Beta}(p;\alpha;\beta) = (n+1)\mathrm{Binom}(k;n;p)$$

Beta distributions also provide a family of prior probability distributions for binomial distributions in Bayesian inference:

$$P(p;\alpha,\beta) = \frac{p^{\alpha-1}(1-p)^{\beta-1}}{B(\alpha,\beta)}.$$

Given a uniform prior, the posterior distribution for the probability of success *p* given *n* independent events with *k* observed successes is a beta distribution.

Computational Methods

Generating Binomial Random Variates

Methods for random number generation where the marginal distribution is a binomial distribution are well-established.

One way to generate random samples from a binomial distribution is to use an inversion algorithm. To do so, one must calculate the probability that Pr(X = k) for all values k from 0 through n. (These probabilities should sum to a value close to one, in order to encompass the entire sample space.) Then by using a pseudorandom number generator to generate samples uniformly between 0 and 1, one can transform the calculated samples into discrete numbers by using the probabilities calculated in the first step.

Hypergeometric Distribution

In probability theory and statistics, the hypergeometric distribution is a discrete probability distribution that describes the probability of k successes (random draws for which the object drawn has a specified feature) in n draws, without replacement, from a finite population of size N that contains exactly K objects with that feature, wherein each draw is either a success or a failure. In contrast, the binomial distribution describes the probability of k successes in n draws with replacement.

In statistics, the hypergeometric test uses the hypergeometric distribution to calculate the statistical significance of having drawn a specific k successes (out of n total draws) from the aforementioned population. The test is often used to identify which sub-populations are over- or under-represented in a sample. This test has a wide range of applications. For example, a marketing group could use the test to understand their customer base by testing a set of known customers for over-representation of various demographic subgroups (e.g., women, people under 30).

Hypergeometric	
Parameters	$N \in \{0,1,2,\ldots\}$ $K \in \{0,1,2,\ldots,N\}$ $n \in \{0,1,2,\ldots,N\}$
Support	$k \in \{\max(0,n+K-N),\ldots,\min(n,K)\}$
pmf	$\dfrac{\dbinom{K}{k}\dbinom{N-K}{n-k}}{\dbinom{N}{n}}$
CDF	$1-\dfrac{\dbinom{n}{k+1}\dbinom{N-n}{K-k-1}}{\dbinom{N}{K}}\,{}_3F_2\left[\begin{matrix}1, k+1-K, k+1-n\\ k+2, N+k+2-K-n\end{matrix};1\right],$ where ${}_pF_q$ is the generalized hypergeometric function

Mean	$n\frac{K}{N}$
Mode	$\left\lceil \frac{(n+1)(K+1)}{N+2} \right\rceil -1, \left\lfloor \frac{(n+1)(K+1)}{N+2} \right\rfloor$
Variance	$\frac{K}{N}\frac{(N-K)}{N}\frac{N-n}{N\;1}$
Skewness	$\frac{(N-2K)(N-1)^{\frac{1}{2}}(N-2n)}{[nK(N-K)(N-n)]^{\frac{1}{2}}(N-2)}$
Ex. kurtosis	$\frac{1}{nK(N-K)(N-n)(N-2)(N-3)} \cdot \left[(N-1)N^2\left(N(N+1)-6K(N-K)-6n(N-n)\right)+6nK(N-K)(N-n)(5N-6)\right]$
MGF	$\frac{\binom{N-K}{n}\,{}_2F_1(-n,-K;N-K-n+1;e^t)}{\binom{N}{n}}$
CF	$\frac{\binom{N-K}{n}\,{}_2F_1(-n,-K;N-K-n+1;e^{it})}{\binom{N}{n}}$

The following conditions characterize the hypergeometric distribution:

- The result of each draw (the elements of the population being sampled) can be classified into one of two mutually exclusive categories (e.g. Pass/Fail or Employed/Unemployed).
- The probability of a success changes on each draw, as each draw decreases the population (sampling without replacement from a finite population).

A random variable X follows the hypergeometric distribution if its probability mass function (pmf) is given by:

$$p_X(k) = \Pr(X=k) = \frac{\binom{K}{k}\binom{N-K}{n-k}}{\binom{N}{n}},$$

where,

- N is the population size,
- K is the number of success states in the population,
- n is the number of draws (i.e. quantity drawn in each trial),
- k is the number of observed successes,
- $\binom{a}{b}$ is a binomial coefficient.

The pmf is positive when $\max(0, n+K-N) \leq k \leq \min(K, n)$.

A random variable distributed hypergeometrically with parameters N, K and n is written $X \sim \text{Hypergeometric}(N, K, n)$ and has probability mass function $p_X(k)$ above.

Combinatorial Identities

As required, we have,

$$\sum_{0 \leq k \leq n} \frac{\binom{K}{k}\binom{N-K}{n-k}}{\binom{N}{n}} = 1,$$

which essentially follows from Vandermonde's identity from combinatorics.

Also note that,

$$\frac{\binom{K}{k}\binom{N-K}{n-k}}{\binom{N}{n}} = \frac{\binom{n}{k}\binom{N-n}{K-k}}{\binom{N}{K}},$$

which follows from the symmetry of the problem, but it can also be shown by expressing the binomial coefficients in terms of factorials and rearranging the latter.

Properties

Working Example

The classical application of the hypergeometric distribution is sampling without replacement. Think of an urn with two types of marbles, red ones and green ones. Define drawing a green marble as a success and drawing a red marble as a failure (analogous to the binomial distribution). If the variable N describes the number of all marbles in the urn and K describes the number of green marbles, then N – K corresponds to the number of red marbles. In this example, X is the random variable whose outcome is k, the number of green marbles actually drawn in the experiment. This situation is illustrated by the following contingency table:

	Drawn	Not drawn	Total
Green marbles	K	K – k	K
Red marbles	N – k	N + k – n – K	N – K
Total	N	N – n	N

Now, assume (for example) that there are 5 green and 45 red marbles in the urn. Standing next to the urn, you close your eyes and draw 10 marbles without replacement. What is the probability that exactly 4 of the 10 are green? Note that although we are looking at success/failure, the data are not accurately modeled by the binomial distribution, because the probability of success on each trial is not the same, as the size of the remaining population changes as we remove each marble.

This problem is summarized by the following contingency table:

	Drawn	Not drawn	Total
Green marbles	K = 4	K – k = 1	K = 5
Red marbles	N – k = 6	N + k – n – K = 39	N – K = 45
Total	N = 10	N – n = 40	N = 50

The probability of drawing exactly *k* green marbles can be calculated by the formula,

$$P(X=k)=f(k;N,K,n)=\frac{\binom{K}{k}\binom{N-K}{n-k}}{\binom{N}{n}}.$$

Hence, in this example calculate,

$$P(X=4)=f(4;50,5,10)=\frac{\binom{5}{4}\binom{45}{6}}{\binom{50}{10}}=\frac{5\cdot 8145060}{10272278170}=0.003964583\ldots$$

Intuitively we would expect it to be even more unlikely that all 5 green marbles will be among the 10 drawn.

$$P(X=5)=f(5;50,5,10)=\frac{\binom{5}{5}\binom{45}{5}}{\binom{50}{10}}=\frac{1\cdot 1221759}{10272278170}=0.0001189375\ldots,$$

As expected, the probability of drawing 5 green marbles is roughly 35 times less likely than that of drawing 4.

Symmetries

Swapping the roles of green and red marbles:

$$f(k;N,K,n)=f(n-k;N,N-K,n)$$

Swapping the roles of drawn and not drawn marbles:

$$f(k;N,K,n)=f(K-k;N,K,N-n)$$

Swapping the roles of green and drawn marbles:

$$f(k;N,K,n)=f(k;N,n,K)$$

Order of Draws

The probability of drawing any set of green and red marbles (the hypergeometric distribution) depends only on the numbers of green and red marbles, not on the order in which they appear; i.e., it is an exchangeable distribution. As a result, the probability of drawing a green marble in the ith draw is,

$$P(G_i)=\frac{K}{N}.$$

This is an ex ante probability—that is, it is based on not knowing the results of the previous draws.

Tail Bounds

Let $X \sim \text{Hypergeometric}(K,N,n)$ and $p=K/N$. Then for $0<t<nK/N$ we can derive the following bounds:

$$\Pr[X\le(p-t)n]\le e^{-nD(p-t\|p)}\le e^{-2t^2n}$$

$$\Pr[X\ge(p+t)n\le e^{-nD(p+t\|p)}\le e^{-2t^2n}$$

where,

$$D(a\|b)=a\log\frac{a}{b}+(1-a)\log\frac{1-a}{1-b}$$

is the Kullback-Leibler divergence and it is used that $D(a\|b)\ge 2(a-b)^2$.

If n is larger than $N/2$, it can be useful to apply symmetry to "invert" the bounds, which give you the following:

$$\Pr[X \le (p-t)n] \le e^{-(N-n)D(p+\frac{tn}{N-n}||p)} \le e^{-2t^2 n \frac{n}{N-n}}$$

$$\Pr[X \ge (p+t)n] \le e^{-(N-n)D(p-\frac{tn}{N-n}||p)} \le e^{-2t^2 n \frac{n}{N-n}}$$

Statistical Inference

Hypergeometric Test

The hypergeometric test uses the hypergeometric distribution to measure the statistical significance of having drawn a sample consisting of a specific number of k successes (out of n total draws) from a population of size N containing K successes. In a test for over-representation of successes in the sample, the hypergeometric p-value is calculated as the probability of randomly drawing k or more successes from the population in n total draws. In a test for under-representation, the p-value is the probability of randomly drawing k or fewer successes.

The test based on the hypergeometric distribution (hypergeometric test) is identical to the corresponding one-tailed version of Fisher's exact test). Reciprocally, the p-value of a two-sided Fisher's exact test can be calculated as the sum of two appropriate hypergeometric tests.

Related Distributions

Let $X \sim \text{Hypergeometric}(N,K,n)$ and $p = K/N$.

- If $n = 1$ then X has a Bernoulli distribution with parameter p.
- Let Y have a binomial distribution with parameters n and p; this models the number of successes in the analogous sampling problem *with* replacement. If N and K are large compared to n, and p is not close to 0 or 1, then X and Y have similar distributions, i.e.,.
- If n is large, N and K are large compared to n, and p is not close to 0 or 1, then,

  $$P(X \le k) \approx \Phi\left(\frac{k-np}{\sqrt{np(1-p)}}\right)$$

  where Φ is the standard normal distribution function,
- If the probabilities of drawing a green or red marble are not equal (e.g. because green marbles are bigger/easier to grasp than red marbles) then X has a noncentral hypergeometric distribution.
- The beta-binomial distribution is a conjugate prior for the hypergeometric distribution.

The following table describes four distributions related to the number of successes in a sequence of draws:

	With replacements	No replacements
Given number of draws	binomial distribution	hypergeometric distribution
Given number of failures	negative binomial distribution	negative hypergeometric distribution

Multivariate Hypergeometric Distribution

The model of an urn with green and red marbles can be extended to the case where there are more than two colors of marbles. If there are *Ki* marbles of color *i* in the urn and you take *n* marbles at random without replacement, then the number of marbles of each color in the sample (k_1,k_2,...,*kc*) has the multivariate hypergeometric distribution. This has the same relationship to the multinomial distribution that the hypergeometric distribution has to the binomial distribution—the multinomial distribution is the "with-replacement" distribution and the multivariate hypergeometric is the "without-replacement" distribution.

Multivariate hypergeometric distribution	
Parameters	$c \in \mathbb{N}_+ = \{1,2,\ldots\}\ (K_1,\ldots,K_c) \in \mathbb{N}^c$ $N = \sum_{i=1}^{c} K_i \quad n \in \{0,\ldots,N\}$
Support	$\left\{\mathbf{k} \in \mathbb{Z}_{0+}^{c} : \forall i\, k_i \le K_i, \sum_{i=1}^{c} k_i = n\right\}$
pmf	$\dfrac{\prod_{i=1}^{c} \binom{K_i}{k_i}}{\binom{N}{n}}$
Mean	$E(X_i) = n\dfrac{K_i}{N}$
Variance	$Var(X_i) = n\dfrac{N-n}{N-1}\dfrac{K_i}{N}\left(1-\dfrac{K_i}{N}\right)$ $Cov(X_i,X_j) = -n\dfrac{N-n}{N-1}\dfrac{K_i}{N}\dfrac{K_j}{N}$

The properties of this distribution are given in the adjacent table, where c is the number of different colors and $N = \sum_{i=1}^{c} K_i$ is the total number of marbles.

Suppose there are 5 black, 10 white, and 15 red marbles in an urn. If six marbles are chosen without replacement, the probability that exactly two of each color are chosen is,

$$P(2\text{ black}, 2\text{ white}, 2\text{ red}) = \frac{\binom{5}{2}\binom{10}{2}\binom{15}{2}}{\binom{30}{6}} = 0.0795755968169 76$$

Occurrence and Applications

Application to Auditing Elections

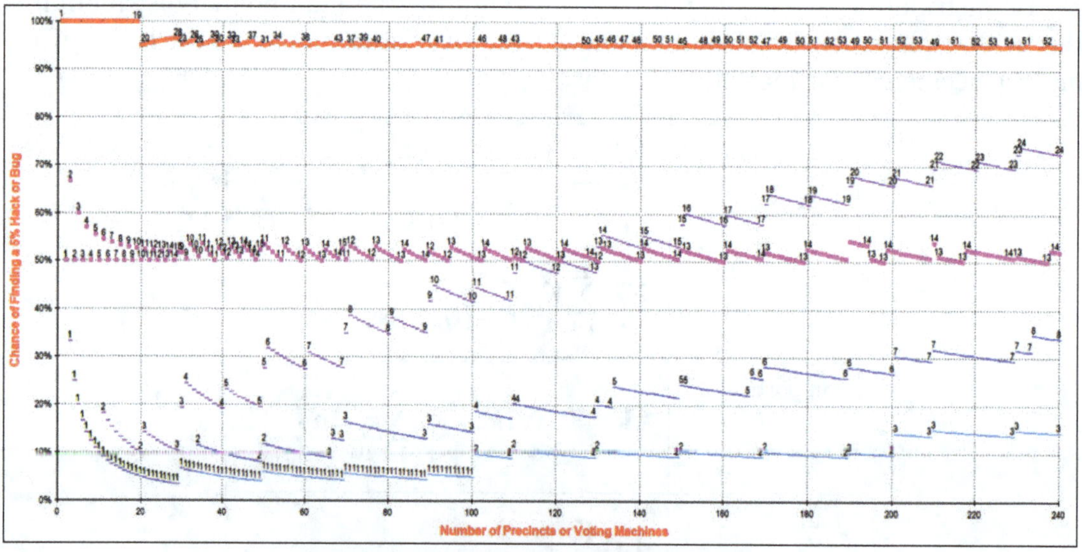

Samples used for election audits and resulting chance of missing a problem.

Election audits typically test a sample of machine-counted precincts to see if recounts by hand or machine match the original counts. Mismatches result in either a report or a larger recount. The sampling rates are usually defined by law, not statistical design, so for a legally defined sample size n, what is the probability of missing a problem which is present in K precincts, such as a hack or bug? This is the probability that $k = 0$. Bugs are often obscure, and a hacker can minimize detection by affecting only a few precincts, which will still affect close elections, so a plausible scenario is for K to be on the order of 5% of N. Audits typically cover 1% to 10% of precincts (often 3%), so they have a high chance of missing a problem. For example, if a problem is present in 5 of 100 precincts, a 3% sample has 86% probability that $k = 0$ so the problem would not be noticed, and only 14% probability of the problem appearing in the sample (positive k):

$$X=0) = \frac{\binom{\text{Hack}}{0}\binom{N-\text{Hack}}{n-0}}{\binom{N}{n}} = \frac{\binom{N-\text{Hack}}{n}}{\binom{N}{n}} = \frac{\frac{(N-\text{Hack})!}{n!(N-\text{Hack}-n)!}}{\frac{N!}{n!(N-n)!}} = \frac{\frac{(N-\text{Hack})!}{(N-\text{Hack}-n)!}}{\frac{N!}{(N-n)!}}$$

$$= \frac{\binom{100-5}{3}}{\binom{100}{3}} = \frac{\frac{(100-5)!}{(100-5-3)!}}{\frac{100!}{(100-3)!}} = \frac{\frac{95!}{92!}}{\frac{100!}{97!}} = \frac{95\times94\times93}{100\times99\times98} = 86\%$$

The sample would need 45 precincts in order to have probability under 5% that $k = 0$ in the sample, and thus have probability over 95% of finding the problem:

$$P(X=0) = \frac{\binom{100-5}{45}}{\binom{100}{45}} = \frac{\frac{95!}{50!}}{\frac{100!}{55!}} = \frac{95\times94\times\cdots\times51}{100\times99\times\cdots\times56} = \frac{55\times54\times53\times52\times51}{100\times99\times98\times97\times96} = 4.6\%$$

Application to Texas Hold'em Poker

In hold'em poker players make the best hand they can combining the two cards in their hand with the 5 cards (community cards) eventually turned up on the table. The deck has 52 and there are 13 of each suit. For this example assume a player has 2 clubs in the hand and there are 3 cards showing on the table, 2 of which are also clubs. The player would like to know the probability of one of the next 2 cards to be shown being a club to complete the flush. (Note that the probability calculated in this example assumes no information is known about the cards in the other players' hands; however, experienced poker players may consider how the other players place their bets (check, call, raise, or fold) in considering the probability for each scenario. Strictly speaking, the approach to calculating success probabilities outlined here is accurate in a scenario where there is just one player at the table; in a multiplayer game this probability might be adjusted somewhat based on the betting play of the opponents.)

There are 4 clubs showing so there are 9 clubs still unseen. There are 5 cards showing (2 in the hand and 3 on the table) so there are $52-5=47$ still unseen.

The probability that one of the next two cards turned is a club can be calculated using hypergeometric with $k=1, n=2, K=9$ and $N=47$. (about 31.6%).

The probability that both of the next two cards turned are clubs can be calculated using hypergeometric with $k=2, n=2, K=9$ and $N=47$. (about 3.3%).

The probability that neither of the next two cards turned are clubs can be calculated using hypergeometric with $k=0, n=2, K=9$ and $N=47$. (about 65.0%).

Poisson Distribution

Given a Poisson process, the probability of obtaining exactly n successes in N trials is given by the limit of a binomial distribution.

$$P_p(n|N)=\frac{N!}{n!(N-n)!}p^n(1-p)^{(N-n)}.$$

Viewing the distribution as a function of the expected number of successes,

$$v \equiv Np$$

instead of the sample size N for fixed p, equation ($v \equiv Np$) then becomes,

$$P_{v/N}(n|N)=\frac{N!}{n!(N-n)!}\left(\frac{v}{N}\right)^n\left(1-\frac{v}{N}\right)^{N-n},$$

Letting the sample size N become large, the distribution then approaches,

$$\begin{aligned}
P_v(n) &= \lim_{N\to\infty} P_p(n|N) \\
&= \lim_{N\to\infty}\frac{N(N-1)\ldots(N-n+1)}{n!}\frac{v^n}{N^n}\left(1-\frac{v}{N}\right)^N\left(1-\frac{v}{N}\right)^{-n} \\
&= \lim_{N\to\infty}\frac{N(N-1)\ldots(N-n+1)}{N^n}\frac{v^n}{n!}\left(1-\frac{v}{N}\right)^N\left(1-\frac{v}{N}\right)^{-n} \\
&= 1.\frac{v^n}{n!}.e^{-v}.1 \\
&= \frac{v^n e^{-v}}{n!}
\end{aligned}$$

which is known as the Poisson distribution. Note that the sample size N has completely dropped out of the probability function, which has the same functional form for all values of v.

As expected, the Poisson distribution is normalized so that the sum of probabilities equals 1, since,

$$\sum_{n=0}^{\infty} P_v(n) = e^{-v}\sum_{n=0}^{\infty}\frac{v^n}{n!} = e^{-v}e^{v} = 1.$$

The ratio of probabilities is given by,

$$\frac{P_v(n=i+1)}{P(n=i)} = \frac{\dfrac{v^{i+1}e^{-v}}{(i+1)!}}{\dfrac{e^{-v}v^{i}}{i!}} = \frac{v}{i+1}$$

The Poisson distribution reaches a maximum when,

$$\frac{dP_v(n)}{dn} = \frac{e^{-v}n(\gamma - H_n + \ln v)}{n!} = 0$$

where γ is the Euler-Mascheroni constant and H_n is a harmonic number, leading to the transcendental equation.

$$\gamma - H_n + \ln v = 0,$$

which cannot be solved exactly for n.

The moment-generating function of the Poisson distribution is given by:

$$\begin{aligned}
M(t) &= e^{(-v)}e^{ve^t} = e^{v(e^t-1)} \\
M'(t) &= ve^te^{v(e^t-1)} \\
M''(t) &= (ve^t)^2e^{v(e^t-1)} + ve^te^{v(e^t-1)} \\
R(t) &= nu(e^t-1) \\
R(t) &= ve^t \\
R''(t) &= ve^t,
\end{aligned}$$

So,

$$\begin{aligned}
\mu &= R'(0) = v \\
\sigma^2 &= R''(0) = v
\end{aligned}$$

The raw moments can also be computed directly by summation, which yields an unexpected connection with the Bell polynomial $\phi_n(x)$ and Stirling numbers of the second kind,

$$\phi_n(x) = \sum_{k=0}^{\infty}\frac{e^{-x}x^k}{k!}k_n = \sum_{k=1}^{n}x^kS(n,k)$$

known as Dobiński's formula. Therefore,

$$\begin{aligned} \mu_2' &= v(1+v) \\ \mu_3' &= v\left(1+3v+v^2\right) \\ \mu_4' &= v\left(1+7v+6v^2+v^3\right) \end{aligned}$$

The central moments can then be computed as,

$$\mu_2 = v$$
$$\mu_3 = v$$
$$\mu_4 = v(1+3v),$$

so the mean, variance, skewness, and kurtosis excess are,

$$\mu = v$$
$$\sigma^2 = v$$
$$\gamma \equiv \frac{\mu_3}{\sigma^3} = \frac{v}{v^{3/2}} = v^{-1/2}$$
$$\gamma \equiv \frac{\mu_4}{\sigma^4} - 3 = \frac{v(1+3v)}{v^2} - 3$$
$$= \frac{v+3v^2-3v^2}{v^2} = v^{-1}$$

The characteristic function for the Poisson distribution is,

$$\phi(t) = e^{v\left(e^{it}-1\right)}$$

and the cumulant-generating function is,

$$K(h) = v\left(e^h - 1\right) = v\left(h + \frac{1}{2!}h^2 + \frac{1}{3!}h^3 + \ldots\right),$$

So, $\kappa_r = v$. The mean deviation of the Poisson distribution is given by,

$$MD = \frac{2e^{-v}v^{\lfloor v \rfloor + 1}}{\lfloor v \rfloor !}$$

The Poisson distribution can also be expressed in terms of,

$$\lambda \equiv \frac{v}{x},$$

the rate of changes, so that,

$$P_v(n) = \frac{(\lambda x)^n e^{-\lambda x}}{n!}$$

The moment-generating function of a Poisson distribution in two variables is given by,

$$M(t) = e^{(v_1+v_2)(e^t-1)}$$

If the independent variables $x_1, x_2, \ldots x_N$ have Poisson distributions with parameters $\mu_1, \mu_2, \ldots \mu_N$ then,

$$X = \sum_{j=1}^{N} x_j$$

has a Poisson distribution with parameter,

$$\mu = \sum_{j=1}^{N} \mu_j$$

This can be seen since the cumulant-generating function is,

$$K_j(h) = \mu_j(e^h - 1)$$

$$K \equiv \sum_j K_j(h) = (e^h - 1)\sum_j \mu_j = \mu(e^h - 1)$$

A generalization of the Poisson distribution has been used by Saslaw to model the observed clustering of galaxies in the universe. The form of this distribution is given by,

$$f_b(N) = \frac{\bar{N}(1-b)}{N!}\left[\bar{N}(1-b) + Nb\right]^{N-1} e^{\bar{N}(1-b)-Nb},$$

where N is the number of galaxies in a volume V, $\bar{N} = \bar{n}V, \bar{n}$ is the average density of galaxies, and $b = -W/(2K) \approx 0.70 \pm 0.05$, with $0 \le b < 1$ is the ratio of gravitational energy to the kinetic energy of peculiar motions, Letting $b = 0$ gives,

$$f_0(N) = \frac{e^{-\bar{N}}\,\bar{N}^N}{N!},$$

which is indeed a Poisson distribution with $v = \bar{N}$. Similarly, letting $b = 1$ gives $f_1(N) = 0$.

Discrete Distributions without Infinite Support

Geometric Distribution

In probability theory and statistics, the geometric distribution is either of two discrete probability distributions:

- The probability distribution of the number *X* of Bernoulli trials needed to get one success, supported on the set { 1, 2, 3, ... }.

- The probability distribution of the number $Y = X - 1$ of failures before the first success, supported on the set { 0, 1, 2, 3, ... }.

Which of these one calls "the" geometric distribution is a matter of convention and convenience.

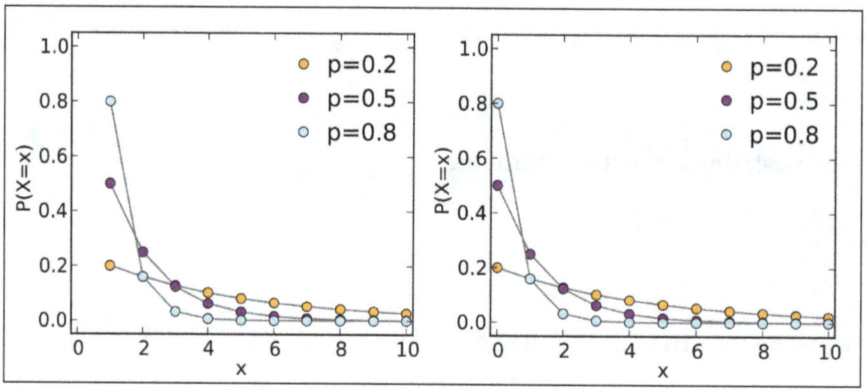

Probability mass function.

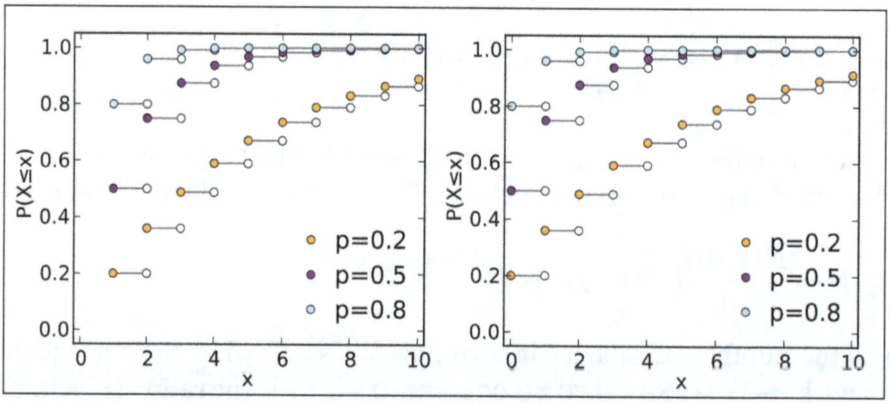

Cumulative distribution function.

These two different geometric distributions should not be confused with each other. Often, the name *shifted* geometric distribution is adopted for the former one (distribution of the number X); however, to avoid ambiguity, it is considered wise to indicate which is intended, by mentioning the support explicitly.

The geometric distribution gives the probability that the first occurrence of success requires k independent trials, each with success probability p. If the probability of success on each trial is p, then the probability that the kth trial (out of k trials) is the first success is:

$$\Pr(X = k) = (1-p)^{k-1} p$$

for $k = 1, 2, 3, ...$

The above form of the geometric distribution is used for modeling the number of trials up to and including the first success. By contrast, the following form of the geometric distribution is used for modeling the number of failures until the first success:

$$\Pr(Y = k) = (1 - p)^k p$$

for $k = 0, 1, 2, 3, \ldots$.

In either case, the sequence of probabilities is a geometric sequence.

For example, suppose an ordinary die is thrown repeatedly until the first time a "1" appears. The probability distribution of the number of times it is thrown is supported on the infinite set { 1, 2, 3, ... } and is a geometric distribution with $p = 1/6$.

The geometric distribution is denoted by Geo(p) where $0 < p \leq 1$. Consider a sequence of trials, where each trial has only two possible outcomes (designated failure and success). The probability of success is assumed to be the same for each trial. In such a sequence of trials, the geometric distribution is useful to model the number of failures before the first success. The distribution gives the probability that there are zero failures before the first success, one failure before the first success, two failures before the first success, and so on.

Assumptions: When is the Geometric Distribution an Appropriate Model

The geometric distribution is an appropriate model if the following assumptions are true:

- The phenomenon being modeled is a sequence of independent trials.
- There are only two possible outcomes for each trial, often designated success or failure.
- The probability of success, p, is the same for every trial.

If these conditions are true, then the geometric random variable Y is the count of the number of failures before the first success. The possible number of failures before the first success is 0, 1, 2, 3, and so on. In the graphs above, this formulation is shown on the right.

An alternative formulation is that the geometric random variable X is the total number of trials up to and including the first success, and the number of failures is $X - 1$. In the graphs above, this formulation is shown on the left.

Probability Outcomes Examples

The general formula to calculate the probability of k failures before the first success, where the probability of success is p and the probability of failure is $q = 1 - p$, is:

$$\Pr(Y = k) = q^k p.$$

for $k = 0, 1, 2, 3, \ldots$.

E1) A doctor is seeking an anti-depressant for a newly diagnosed patient. Suppose that, of the available anti-depressant drugs, the probability that any particular drug will be effective for a particular patient is $p = 0.6$. What is the probability that the first drug found to be effective for this patient is the first drug tried, the second drug tried, and so on? What is the expected number of drugs that will be tried to find one that is effective?

The probability that the first drug works. There are zero failures before the first success. $Y = 0$ failures. The probability P(zero failures before first success) is simply the probability that the first drug works.

$$\Pr(Y = 0) = q^0 p = 0.4^0 \times 0.6 = 1 \times 0.6 = 0.6.$$

The probability that the first drug fails, but the second drug works. There is one failure before the first success. Y= 1 failure. The probability for this sequence of events is P(first drug fails) p(second drug is success) which is given by,

$$\Pr(Y = 1) = q^1 p = 0.4^1 \times 0.6 = 0.4 \times 0.6 = 0.24.$$

The probability that the first drug fails, the second drug fails, but the third drug works. There are two failures before the first success. $Y = 2$ failures. The probability for this sequence of events is P(first drug fails) × p(second drug fails) × P(third drug is success):

$$\Pr(Y = 2) = q^2 p, = 0.4^2 \times 0.6 = 0.096.$$

E2) A newlywed couple plans to have children, and will continue until the first girl. What is the probability that there are zero boys before the first girl, one boy before the first girl, two boys before the first girl, and so on?

The probability of having a girl (success) is p= 0.5 and the probability of having a boy (failure) is $q = 1 - p = 0.5$.

The probability of no boys before the first girl is,

$$\Pr(Y = 0) = q^0 p = 0.5^0 \times 0.5 = 1 \times 0.5 = 0.5.$$

The probability of one boy before the first girl is,

$$\Pr(Y = 1) = q^1 p = 0.5^1 \times 0.5 = 0.5 \times 0.5 = 0.25.$$

The probability of two boys before the first girl is,

$$\Pr(Y = 2) = q^2 p = 0.5^2 \times 0.5 = 0.125 \text{ and so on.}$$

Properties

Moments and Cumulants

The expected value for the number of independent trials to get the first success, of a geometrically distributed random variable X is $1/p$ and the variance is $(1 - p)/p^2$:

$$E(X) = \frac{1}{p}, \qquad var(X) = \frac{1-p}{p^2}.$$

Similarly, the expected value of the geometrically distributed random variable $Y = X - 1$ (where Y corresponds to the pmf listed in the right column) is $q/p = (1 - p)/p$, and its variance is $(1 - p)/p^2$:

$$E(Y) = \frac{1-p}{p}, \qquad var(Y) = \frac{1-p}{p^2}.$$

Let $\mu = (1 - p)/p$ be the expected value of Y. Then the cumulants κ_n of the probability distribution of Y satisfy the recursion:

$$\kappa_{n+1} = \mu(\mu+1)\frac{d\kappa_n}{d\mu}.$$

Outline of proof: That the expected value is $(1 - p)/p$ can be shown in the following way. Let Y be as above. Then:

$$\begin{aligned} E(Y) &= \sum_{k=0}^{\infty}(1-p)^k p\cdot k \\ &= p\sum_{k=0}^{\infty}(1-p)^k k \\ &= p(1-p)\sum_{k=0}^{\infty}(1-p)^{k-1}\cdot k \\ &= p(1-p)\left[\frac{d}{dp}\left(-\sum_{k=0}^{\infty}(1-p)^k\right)\right] \\ &= p(1-p)\frac{d}{dp}\left(-\frac{1}{p}\right) = \frac{1-p}{p}. \end{aligned}$$

(The interchange of summation and differentiation is justified by the fact that convergent power series converge uniformly on compact subsets of the set of points where they converge).

Expected Value Examples

E3) A patient is waiting for a suitable matching kidney donor for a transplant. If the probability that a randomly selected donor is a suitable match is p=0.1, what is the expected number of donors who will be tested before a matching donor is found?

With $p = 0.1$, the mean number of failures before the first success is $E(Y) = (1 - p)/p = (1 - 0.1)/0.1 = 9$.

For the alternative formulation, where X is the number of trials up to and including the first success, the expected value is $E(X) = 1/p = 1/0.1 = 10$.

For example 1 above, with $p = 0.6$, the mean number of failures before the first success is $E(Y) = (1 - p)/p = (1 - 0.6)/0.6 = 0.67$.

General Properties

- The probability-generating functions of X and Y are, respectively,

$$G_X(s) = \frac{sp}{1 - s(1-p)},$$

$$G_Y(s) = \frac{p}{1 - s(1-p)}, \quad |s| < (1-p)^{-1}.$$

- Like its continuous analogue (the exponential distribution), the geometric distribution is memoryless. That means that if you intend to repeat an experiment until the first success, then, given that the first success has not yet occurred, the conditional probability distribution of the number of additional trials does not depend on how many failures have been observed. The die one throws or the coin one tosses does not have a "memory" of these failures. The geometric distribution is the only memoryless discrete distribution.

$$\Pr\{X > m + n \mid X > n\} = \Pr\{X > m\}$$

  Among all discrete probability distributions supported on $\{1, 2, 3, \ldots\}$ with given expected value μ, the geometric distribution X with parameter $p = 1/\mu$ is the one with the largest entropy.

- The geometric distribution of the number Y of failures before the first success is infinitely divisible, i.e., for any positive integer n, there exist independent identically distributed random variables $Y_1, \ldots, Y_n$ whose sum has the same distribution that Y has. These will not be geometrically distributed unless $n = 1$; they follow a negative binomial distribution.

- The decimal digits of the geometrically distributed random variable Y are a sequence of independent (and *not* identically distributed) random variables. For example, the hundreds digit D has this probability distribution:

$$\Pr(D = d) = \frac{q^{100d}}{1 + q^{100} + q^{200} + \cdots + q^{900}},$$

  where $q = 1 - p$, and similarly for the other digits, and, more generally, similarly for numeral systems with other bases than 10. When the base is 2, this shows that a geometrically distributed random variable can be written as a sum of independent random variables whose probability distributions are indecomposable.

- Golomb coding is the optimal prefix code for the geometric discrete distribution.
- The sum of two independent *Geo*(p) distributed random variables is not a geometric distribution.

Related Distributions

- The geometric distribution Y is a special case of the negative binomial distribution, with $r = 1$. More generally, if $Y_1, ..., Yr$ are independent geometrically distributed variables with parameter p, then the sum:

$$Z = \sum_{m=1}^{r} Y_m$$

follows a negative binomial distribution with parameters r and p.

- The geometric distribution is a special case of discrete compound Poisson distribution.
- If $Y_1, ..., Yr$ are independent geometrically distributed variables (with possibly different success parameters pm), then their minimum,

$$W = \min_{m \in 1,\ldots,r} Y_m$$

is also geometrically distributed, with parameter $p = 1 - \prod_m (1 - p_m)$.

- Suppose $0 < r < 1$, and for $k = 1, 2, 3, ...$ the random variable Xk has a Poisson distribution with expected value $r k/k$. Then:

$$\sum_{k=1}^{\infty} kX_k$$

has a geometric distribution taking values in the set {0, 1, 2, ...}, with expected value $r/(1 - r)$.

- The exponential distribution is the continuous analogue of the geometric distribution. If X is an exponentially distributed random variable with parameter λ, then:

$$Y = \lfloor X \rfloor$$

where $\lfloor\ \rfloor$ is the floor (or greatest integer) function, is a geometrically distributed random variable with parameter $p = 1 - e^{-}\lambda$ (thus $\lambda = -\ln(1 - p)$) and taking values in the set {0, 1, 2, ...}. This can be used to generate geometrically distributed pseudorandom numbers by first generating exponentially distributed pseudorandom numbers from a uniform pseudorandom number generator: then $\lfloor \ln(U)/\ln(1-p) \rfloor$ is geometrically distributed with parameter p, if U is uniformly distributed in [0,1].

- If $p = 1/n$ and X is geometrically distributed with parameter p, then the distribution of X/n approaches an exponential distribution with expected value 1 as $n \to \infty$, since:

$$P(X/n > a) = P(X > na) = (1-p)^{na} = \left(1-\frac{1}{n}\right)^{na} = \left[\left(1-\frac{1}{n}\right)^{n}\right]^{a}$$

$$\to [e^{-1}]^{a} = e^{-a} \text{ as } n \to \infty.$$

More generally, if p=λx/n, where λ is a parameter, then as n→ ∞ the distribution approaches an exponential distribution with expected value λ which gives the general definition of the exponential distribution.

$$P(X > x) = \lim_{n\to\infty}(1-\lambda x/n)^{n} = \lambda e^{-\lambda x}$$

therefore the distribution function of x equals $1-e^{-\lambda x}$ and differentiating the probability density function of the exponential function is obtained $f_X(x) = \lambda e^{-\lambda x}$ for $x \geq 0$.

Statistical Inference

Parameter Estimation

For both variants of the geometric distribution, the parameter p can be estimated by equating the expected value with the sample mean. This is the method of moments, which in this case happens to yield maximum likelihood estimates of p.

Specifically, for the first variant let $k = k_1, ..., kn$ be a sample where $ki \geq 1$ for $i = 1, ..., n$. Then p can be estimated as,

$$\hat{p} = \left(\frac{1}{n}\sum_{i=1}^{n} k_i\right)^{-1} = \frac{n}{\sum_{i=1}^{n} k_i}.$$

In Bayesian inference, the Beta distribution is the conjugate prior distribution for the parameter p. If this parameter is given a Beta(α, β) prior, then the posterior distribution is,

$$p \sim \text{Beta}\left(\alpha + n, \beta + \sum_{i=1}^{n}(k_i - 1)\right).$$

The posterior mean E[p] approaches the maximum likelihood estimate $\hat{p}$ as α and β approach zero.

In the alternative case, let k_1, ..., kn be a sample where $ki \geq 0$ for $i = 1, ..., n$. Then p can be estimated as,

$$\hat{p} = \left(1 + \frac{1}{n}\sum_{i=1}^{n} k_i\right)^{-1} = \frac{n}{\sum_{i=1}^{n} k_i + n}.$$

The posterior distribution of p given a Beta(α, β) prior is,

$$p \sim \text{Beta}\left(\alpha + n, \beta + \sum_{i=1}^{n} k_i\right).$$

Again the posterior mean E[p] approaches the maximum likelihood estimate ^ as α and β approach zero.

For either estimate of $\hat{p}$ using Maximum Likelihood, the bias is equal to,

$$b \equiv E\left[(\hat{p}_{mle} - p)\right] = \frac{p(1-p)}{n}$$

which yields the bias-corrected maximum likelihood estimator,

$$\hat{p}^{*}_{mle} = \hat{p}_{mle} - \hat{b}$$

Computational Methods

Geometric Distribution using R

The R function dgeom(k, prob) calculates the probability that there are k failures before the first success, where the argument “prob” is the probability of success on each trial.

For example,

dgeom(0,0.6) = 0.6

dgeom(1,0.6) = 0.24

R uses the convention that k is the number of failures, so that the number of trials up to and including the first success is $k + 1$.

The following R code creates a graph of the geometric distribution from $Y = 0$ to 10, with $p = 0.6$.

```
Y=0:10
```

```
plot(Y, dgeom(Y,0.6), type=”h”, ylim=c(0,1), main=”Geometric distribution for p=0.6”, ylab=”P(Y=Y)”, xlab=”Y=Number of failures before first success”).
```

Geometric Distribution using Excel

The geometric distribution, for the number of failures before the first success, is a special case of the negative binomial distribution, for the number of failures before s successes.

The Excel function NEGBINOMDIST (number_f, number_s, probability_s) calculates the probability of k = number_f failures before s = number_s successes where p = probability_s is the probability of success on each trial. For the geometric distribution, let number_s = 1 success.

For example,

= NEGBINOMDIST (0, 1, 0.6) = 0.6

= NEGBINOMDIST (1, 1, 0.6) = 0.24

Like R, Excel uses the convention that k is the number of failures, so that the number of trials up to and including the first success is k + 1.

Borel Distribution

The Borel distribution is a discrete probability distribution, arising in contexts including branching processes and queueing theory. It is named after the French mathematician Émile Borel.

Borel distribution	
Parameters	$\mu \in [0,1]$
Support	$n \in \{1,2,3,\ldots\}$
pmf	$\frac{e^{-\mu n}(\mu n)^{n-1}}{n!}$
Mean	$\frac{1}{1-\grave{\imath}}$
Variance	$\frac{\grave{\imath}}{(1-\grave{\imath})^3}$

If the number of offspring that an org anism has is Poisson-distributed, and if the average number of offspring of each organism is no bigger than 1, then the descendants of each individual will ultimately become extinct. The number of descendants that an individual ultimately has in that situation is a random variable distributed according to a Borel distribution.

A discrete random variable X is said to have a Borel distribution with parameter $\mu \in [0,1]$ if the probability mass function of X is given by,

$$P_\mu(n) = \Pr(X = n) = \frac{e^{-\mu n}(\mu n)^{n-1}}{n!}$$

for $n = 1, 2, 3 ....$

Derivation and Branching Process Interpretation

If a Galton–Watson branching process has common offspring distribution Poisson with mean μ, then the total number of individuals in the branching process has Borel distribution with parameter μ.

Let X be the total number of individuals in a Galton–Watson branching process. Then a correspondence between the total size of the branching process and a hitting time for an associated random walk gives:

$$\Pr(X = n) = \frac{1}{n}\Pr(S_n = n - 1)$$

where $Sn = Y_1 + ... + Yn$, and $Y_1 ... Yn$ are independent identically distributed random variables whose common distribution is the offspring distribution of the branching process. In the case where this common distribution is Poisson with mean μ, the random variable Sn has Poisson distribution with mean μn, leading to the mass function of the Borel distribution given above.

Since the mth generation of the branching process has mean size μm^{-1}, the mean of X is:

$$1 + \mu + \mu^2 + \cdots = \frac{1}{1-\mu}.$$

Queueing Theory Interpretation

In an M/D/1 queue with arrival rate μ and common service time 1, the distribution of a typical busy period of the queue is Borel with parameter μ.

Properties

If $P\mu(n)$ is the probability mass function of a Borel(μ) random variable, then the mass function $P^* \mu(n)$ of a sized-biased sample from the distribution (i.e. the mass function proportional to $nP\mu(n)$) is given by,

$$P_\mu^*(n) = (1-\mu)\frac{e^{-\mu n}(\mu n)^{n-1}}{(n-1)!}.$$

Aldous and Pitman show that,

$$P_\mu(n) = \frac{1}{\mu}\int_0^\mu P_\lambda^*(n)d\lambda.$$

In words, this says that a Borel(μ) random variable has the same distribution as a size-biased Borel(μU) random variable, where U has the uniform distribution on [0,1].

This relation leads to various useful formulas, including,

$$E\left(\frac{1}{X}\right) = 1 - \frac{\mu}{2}.$$

Borel–Tanner Distribution

The Borel–Tanner distribution generalizes the Borel distribution. Let k be a positive integer. If $X_1, X_2, \ldots Xk$ are independent and each has Borel distribution with parameter μ, then their sum $W = X_1 + X_2 + \ldots + Xk$ is said to have Borel–Tanner distribution with parameters μ and k. This gives the distribution of the total number of individuals in a Poisson–Galton–Watson process starting with k individuals in the first generation, or of the time taken for an M/D/1 queue to empty starting with k jobs in the queue. The case $k = 1$ is simply the Borel distribution above.

Generalizing the random walk correspondence given above for $k = 1$,

$$\Pr(W = n) = \frac{k}{n}\Pr(S_n = n - k)$$

where Sn has Poisson distribution with mean $n\mu$. As a result, the probability mass function is given by,

$$\Pr(W = n) = \frac{k}{n}\frac{e^{-\mu n}(\mu n)^{n-k}}{(n-k)!}$$

for $n = k, k + 1, \ldots,$.

Beta Negative Binomial Distribution

In probability theory, a beta negative binomial distribution is the probability distribution of a discrete random variable X equal to the number of failures needed to get r successes in a sequence of independent Bernoulli trials where the probability p of success on each trial is constant within any given experiment but is itself a random variable following a beta distribution, varying between different experiments. Thus the distribution is a compound probability distribution.

Beta Negative Binomial	
Parameters	shape (real) shape (real) — number of failures until the experiment is stopped (integer but can be extended to real)

Support	$k \in \{0, 1, 2, 3, \ldots\}$
pmf	$\frac{\Gamma(r+k)}{k!\,\Gamma(r)} \frac{B(\alpha+r,\beta+k)}{B(\alpha,\beta)}$
Mean	$\begin{cases} \frac{r\beta}{\alpha-1} & \text{if } \alpha>1 \\ \infty & \text{otherwise} \end{cases}$
Variance	$\begin{cases} \frac{r(\alpha+r-1)\beta(\alpha+\beta-1)}{(\alpha-2)(\alpha-1)^2} & \text{if } \alpha>2 \\ \infty & \text{otherwise} \end{cases}$
Skewness	$\begin{cases} \frac{(\alpha+2r-1)(\alpha+2\beta-1)}{(\alpha-3)\sqrt{\frac{r(\alpha+r-1)\beta(\alpha+\beta-1)}{\alpha-2}}} & \text{if } \alpha>3 \\ \infty & \text{otherwise} \end{cases}$
MGF	undefined
CF	$\frac{B(\alpha,\beta+r)}{B(\alpha,\beta)}\,{}_2F_1(r,\alpha;\alpha+\beta+r;e^{it})$ where B is the beta function and ${}_2F_1$ is the hypergeometric function.

This distribution has also been called both the inverse Markov-Pólya distribution and the generalized Waring distribution. A shifted form of the distribution has been called the beta-Pascal distribution.

If parameters of the beta distribution are α and β, and if,

$$X \mid p \sim NB(r,p),$$

Where,

$$p \sim B(\alpha,\beta),$$

then the marginal distribution of X is a beta negative binomial distribution:

$$X \sim BNB(r,\alpha,\beta).$$

In the above, NB(r, p) is the negative binomial distribution and B(α, β) is the beta distribution.

If r is an integer, then the PMF can be written in terms of the beta function:

$$f(k \mid \alpha,\beta,r) = \binom{r+k-1}{k} \frac{B(\alpha+r,\beta+k)}{B(\alpha,\beta)}.$$

More generally the PMF can be written:

$$f(k \mid \alpha,\beta,r)=\frac{\Gamma(r+k)}{k!\Gamma(r)}\frac{B(\alpha+r,\beta+k)}{B(\alpha,\beta)}.$$

PMF Expressed with Gamma

Using the properties of the Beta function, the PMF with integer r can be rewritten as:

$$f(k \mid \alpha,\beta,r)=\binom{r+k-1}{k}\frac{\Gamma(\alpha+r)\Gamma(\beta+k)\Gamma(\alpha+\beta)}{\Gamma(\alpha+r+\beta+k)\Gamma(\alpha)\Gamma(\beta)}.$$

More generally, the PMF can be written as:

$$f(k \mid \alpha,\beta,r)=\frac{\Gamma(r+k)}{k!\Gamma(r)}\frac{\Gamma(\alpha+r)\Gamma(\beta+k)\Gamma(\alpha+\beta)}{\Gamma(\alpha+r+\beta+k)\Gamma(\alpha)\Gamma(\beta)}.$$

PMF Expressed with the Rising Pochammer Symbol

The PMF is often also presented in terms of the Pochammer symbol for integer r,

$$f(k \mid \alpha,\beta,r)=\frac{r^{(k)}\alpha^{(r)}\beta^{(k)}}{k!(\alpha+\beta)^{(r)}(r+\alpha+\beta)^{(k)}}$$

Properties

The beta negative binomial distribution contains the beta geometric distribution as a special case when $r=1$. It can therefore approximate the geometric distribution arbitrarily well. It also approximates the negative binomial distribution arbitrary well for large *á* and β. It can therefore approximate the Poisson distribution arbitrarily well for large *á*, β and r.

By Stirling's approximation to the beta function, it can be easily shown that,

$$f(k \mid \alpha,\beta,r)\sim\frac{\Gamma(\alpha+r)}{\Gamma(r)B(\alpha,\beta)}\frac{k^{r-1}}{(\beta+k)^{r+\alpha}}$$

which implies that the beta negative binomial distribution is heavy tailed.

Extended Negative Binomial Distribution

In probability and statistics the extended negative binomial distribution is a discrete probability distribution extending the negative binomial distribution. It is a truncated version of the negative binomial distribution for which estimation methods have been studied.

In the context of actuarial science, the distribution appeared in its general form in a paper by K. Hess, A. Liewald and K.D. Schmidt when they characterized all distributions for which the extended Panjer recursion works. For the case m = 1, the distribution was already discussed by Willmot and put into a parametrized family with the logarithmic distribution and the negative binomial distribution by H.U. Gerber.

For a natural number $m \geq 1$ and real parameters p, r with $0 < p \leq 1$ and $-m < r < -m + 1$, the probability mass function of the ExtNegBin(m, r, p) distribution is given by,

$$f(k;m,r,p) = 0 \qquad \text{for } k \in \{0,1,\dots,m-1\}$$

and,

$$f(k;m,r,p) = \frac{\binom{k+r-1}{k} p^k}{(1-p)^{-r} - \sum_{j=0}^{m-1} \binom{j+r-1}{j} p^j} \qquad \text{for } k \in \mathbb{N} \text{ with } k \geq m,$$

where,

$$\binom{k+r-1}{k} = \frac{\Gamma(k+r)}{k!\Gamma(r)} = (-1)^k \binom{-r}{k}$$

is the (generalized) binomial coefficient and Γ denotes the gamma function.

Probability Generating Function

Using that $f(.;m,r,ps)$ for $s \in (0,1]$ is also a probability mass function, it follows that the probability generating function is given by,

$$\begin{aligned}\varphi(s) &= \sum_{k=m}^{\infty} f(k;m,r,p)s^k \\ &= \frac{(1-ps)^{-r} - \sum_{j=0}^{m-1} \binom{j+r-1}{j} (ps)^j}{(1-p)^{-r} - \sum_{j=0}^{m-1} \binom{j+r-1}{j} p^j} \qquad \text{for } |s| \leq \frac{1}{p}.\end{aligned}$$

For the important case $m = 1$, hence $r \in (-1, 0)$, this simplifies to,

$$\varphi(s) = \frac{1-(1-ps)^{-r}}{1-(1-p)^{-r}} \qquad \text{for } |s| \leq \frac{1}{p}.$$

Continuous Distributions

Continuous probability distribution is a A probability distribution in which the random variable X can take on any value (is continuous). Because there are infinite values that

X could assume, the probability of X taking on any one specific value is zero. Therefore we often speak in ranges of values (p(X>0) = .50). The normal distribution is one example of a continuous distribution. The probability that X falls between two values (a and b) equals the integral (area under the curve) from a to b:

Probability Density Function

$$F(x) = p(a \leq x \leq b) = \int_a^b f(x)\,dx \geq 0$$

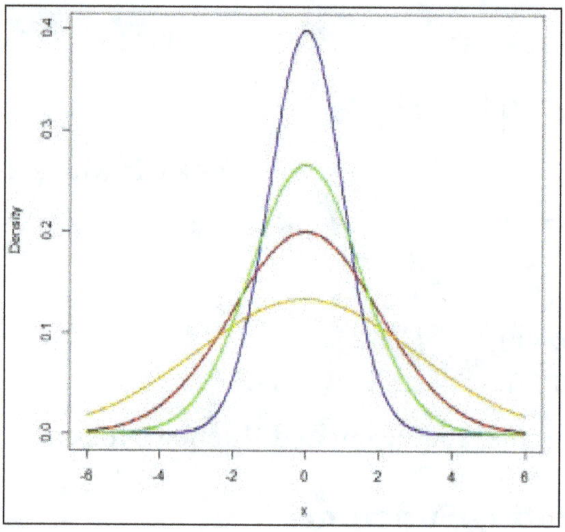

A continuous probability distribution differs from a discrete probability distribution in several ways:

- The probability that a continuous random variable will assume a particular value is zero.
- As a result, a continuous probability distribution cannot be expressed in tabular form.
- Instead, an equation or formula is used to describe a continuous probability distribution.

Beta Distribution

In probability theory and statistics, the beta distribution is a family of continuous probability distributions defined on the interval [0, 1] parametrized by two positive shape parameters, denoted by α and β, that appear as exponents of the random variable and control the shape of the distribution. The generalization to multiple variables is called a Dirichlet distribution.

The beta distribution has been applied to model the behavior of random variables limited to intervals of finite length in a wide variety of disciplines.

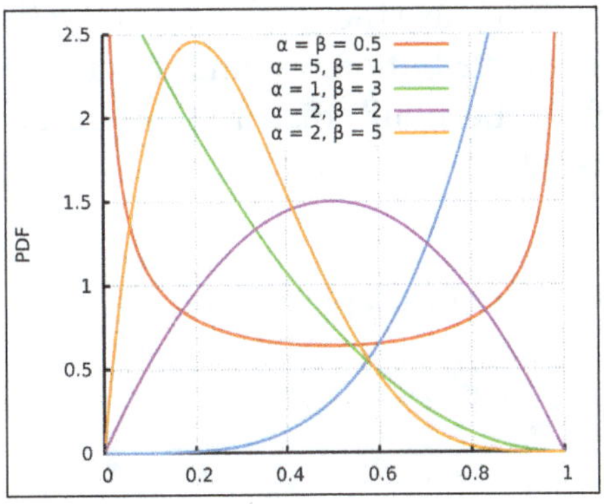

Probability density function.

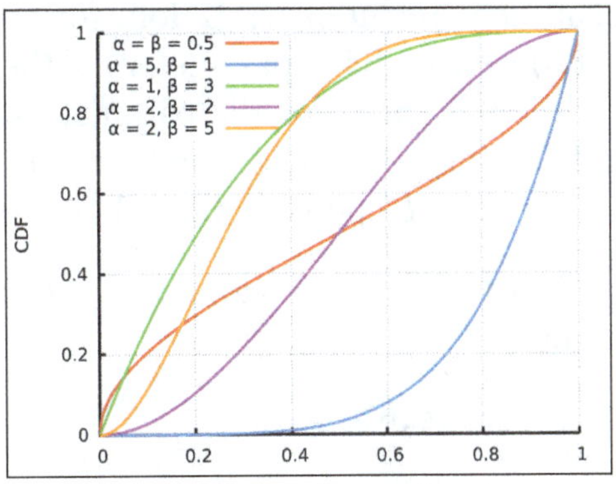

Cumulative distribution function.

In Bayesian inference, the beta distribution is the conjugate prior probability distribution for the Bernoulli, binomial, negative binomial and geometric distributions. For example, the beta distribution can be used in Bayesian analysis to describe initial knowledge concerning probability of success such as the probability that a space vehicle will successfully complete a specified mission. The beta distribution is a suitable model for the random behavior of percentages and proportions.

The probability density function (pdf) of the beta distribution, for $0 \le x \le 1$, and shape parameters $\alpha, \beta > 0$, is a power function of the variable x and of its reflection $(1 - x)$ as follows:

$$
\begin{aligned}
f(x;\alpha,\beta) &= \text{constant}\cdot x^{\alpha-1}(1-x)^{\beta-1} \\
&= \frac{x^{\alpha-1}(1-x)^{\beta-1}}{\int_0^1 u^{\alpha-1}(1-u)^{\beta-1}\,du} \\
&= \frac{\Gamma(\alpha+\beta)}{\Gamma(\alpha)\Gamma(\beta)}x^{\alpha-1}(1-x)^{\beta-1} \\
&= \frac{1}{B(\alpha,\beta)}x^{\alpha-1}(1-x)^{\beta-1}
\end{aligned}
$$

where $\Gamma(z)$ is the gamma function. The beta function, B, is a normalization constant to ensure that the total probability is 1. In the above equations x is a realization—an observed value that actually occurred—of a random process X.

This definition includes both ends $x = 0$ and $x = 1$, which is consistent with definitions for other continuous distributions supported on a bounded interval which are special cases of the beta distribution, for example the arcsine distribution, and consistent with several authors, like N. L. Johnson and S. Kotz. However, the inclusion of $x = 0$ and $x = 1$ does not work for $\alpha, \beta < 1$; accordingly, several other authors, including W. Feller, choose to exclude the ends $x = 0$ and $x = 1$, (so that the two ends are not actually part of the domain of the density function) and consider instead $0 < x < 1$.

Several authors, including N. L. Johnson and S. Kotz, use the symbols p and q (instead of α and β) for the shape parameters of the beta distribution, reminiscent of the symbols traditionally used for the parameters of the Bernoulli distribution, because the beta distribution approaches the Bernoulli distribution in the limit when both shape parameters α and β approach the value of zero.

In the following, a random variable X beta-distributed with parameters α and β will be denoted by:

$$X \sim \mathrm{Beta}(\alpha,\beta)$$

Other notations for beta-distributed random variables used in the statistical literature are $X \sim \mathcal{B}e(\alpha,\beta)$ and $X \sim \beta_{\alpha,\beta}$.

Cumulative Distribution Function

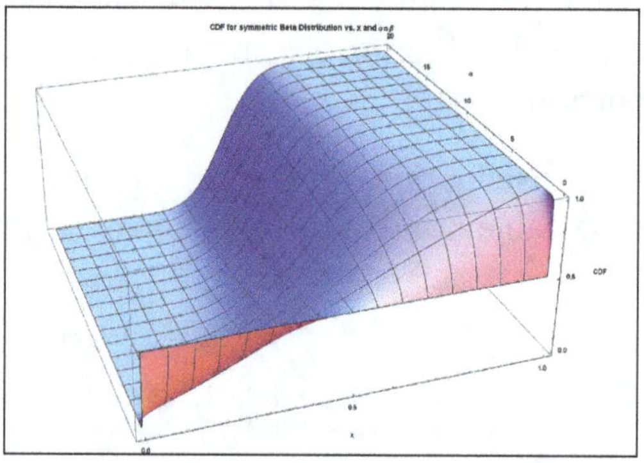

CDF for symmetric beta distribution vs. x and $\alpha = \beta$.

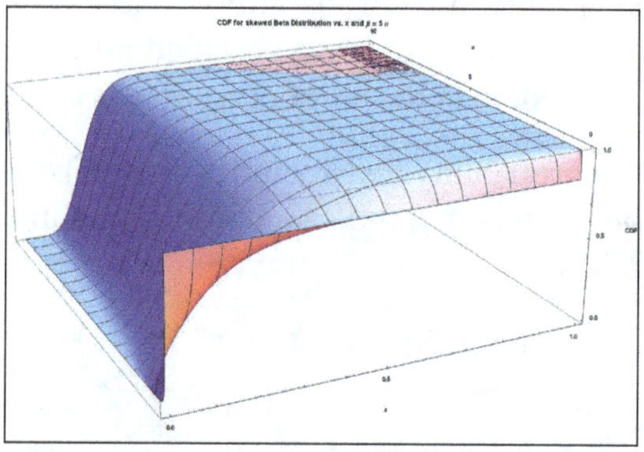

CDF for skewed beta distribution vs. x and $\beta = 5\alpha$.

The cumulative distribution function is,

$$F(x;\alpha,\beta)\frac{B(x;\alpha,\beta)}{B(\alpha,\beta)} = I_x(\alpha,\beta)$$

where $B(x;\alpha,\beta)$ is the incomplete beta function and $I_x(\alpha,\beta)$ is the regularized incomplete beta function.

Alternative Parametrizations and Two Parameters

Mean and Sample Size

The beta distribution may also be reparameterized in terms of its mean μ ($0 < \mu < 1$) and the addition of both shape parameters $\nu = \alpha + \beta > 0$(p. 83). Denoting by αPosterior and βPosterior the shape parameters of the posterior beta distribution resulting from applying Bayes theorem to a binomial likelihood function and a prior probability, the interpretation of the addition of both shape parameters to be sample size = ν =

α·Posterior + β·Posterior is only correct for the Haldane prior probability Beta(0,0). Specifically, for the Bayes (uniform) prior Beta(1,1) the correct interpretation would be sample size = α·Posterior + β Posterior − 2, or ν = (sample size) + 2. Of course, for sample size much larger than 2, the difference between these two priors becomes negligible. $\nu = \alpha + \beta$ will be referred to as "sample size", but one should remember that it is, strictly speaking, the "sample size" of a binomial likelihood function only when using a Haldane Beta(0,0) prior in Bayes theorem.

This parametrization may be useful in Bayesian parameter estimation. For example, one may administer a test to a number of individuals. If it is assumed that each person's score ($0 \le \theta \le 1$) is drawn from a population-level Beta distribution, then an important statistic is the mean of this population-level distribution. The mean and sample size parameters are related to the shape parameters α and β via,

$$\alpha = \mu\nu, \beta = (1 - \mu)\nu$$

Under this parametrization, one may place an uninformative prior probability over the mean, and a vague prior probability (such as an exponential or gamma distribution) over the positive reals for the sample size, if they are independent, and prior data and/or beliefs justify it.

Mode and Concentration

The mode and "concentration" $\kappa = \alpha + \beta$ can also be used to calculate the parameters for a beta distribution.

$$\begin{aligned} \alpha &= \omega(\kappa - 2) + 1 \\ \beta &= (1 - \omega)(\kappa - 2) + 1 \end{aligned}$$

Mean (Allele Frequency) and (Wright's) Genetic Distance between Two Populations

The Balding–Nichols model is a two-parameter parametrization of the beta distribution used in population genetics. It is a statistical description of the allele frequencies in the components of a sub-divided population:

$$\alpha = \mu\nu,$$
$$\beta = (1 - \mu)\nu,$$

where $\nu = \alpha + \beta = \frac{1-F}{F}$ and $0 < F < 1$; here F is (Wright's) genetic distance between two populations.

Mean and Variance

Solving the system of (coupled) equations given in the above sections as the equations for the mean and the variance of the beta distribution in terms of the original parameters α and β, one can express the α and β parameters in terms of the mean (μ) and the variance (var):

$$\nu = \alpha + \beta = \frac{\mu(1-\mu)}{\text{var}} - 1, \text{ where } \nu = (\alpha+\beta) > 0, \text{ therefore: var} < \mu(1-\mu)$$

$$\alpha = \mu\nu = \mu\left(\frac{\mu(1-\mu)}{\text{var}} - 1\right), \text{ if var} < \mu(1-\mu)$$

$$\beta = (1-\mu)\nu = (1-\mu)\left(\frac{\mu(1-\mu)}{\text{var}} - 1\right), \text{ if var} < \mu(1-\mu).$$

This parametrization of the beta distribution may lead to a more intuitive understanding than the one based on the original parameters α and β. For example, by expressing the mode, skewness, excess kurtosis and differential entropy in terms of the mean and the variance:

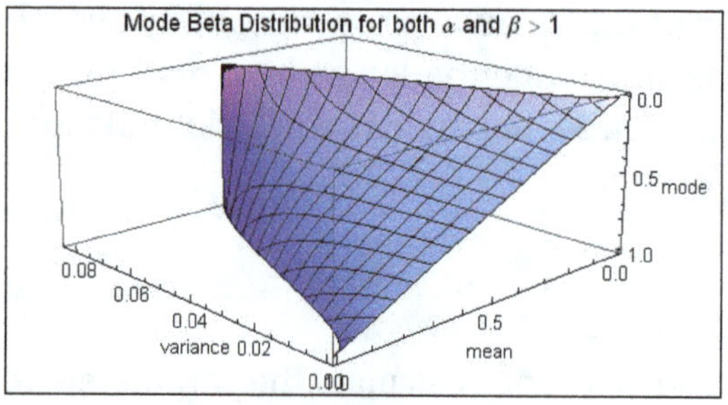

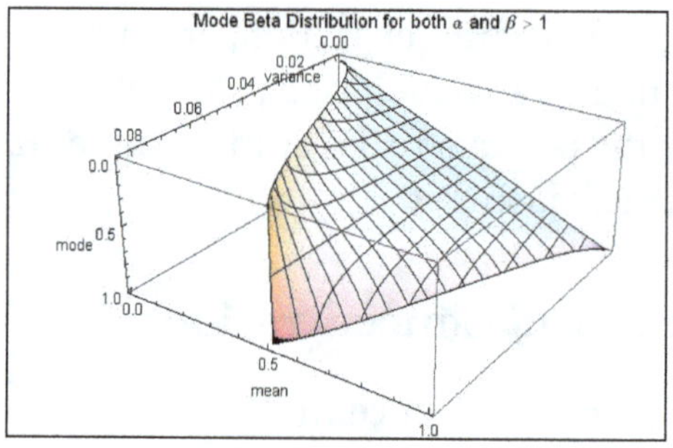

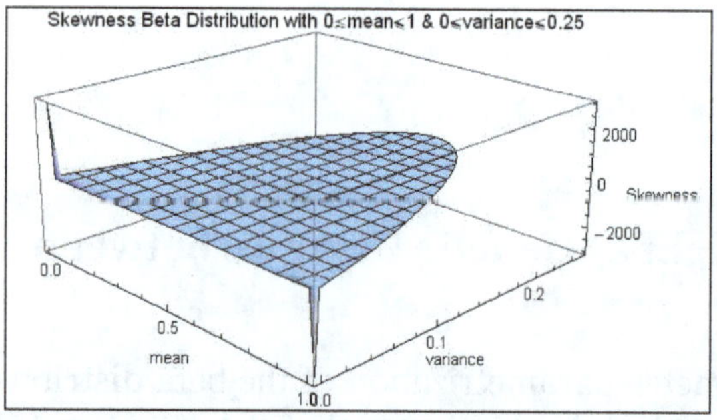

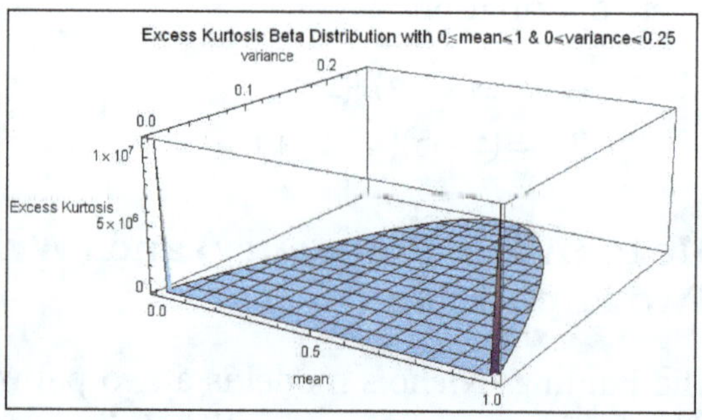

Four Parameters

A beta distribution with the two shape parameters α and β is supported on the range [0,1] or (0,1). It is possible to alter the location and scale of the distribution by introducing two further parameters representing the minimum, *a*, and maximum *c* (*c* > *a*), values of the distribution, by a linear transformation substituting the non-dimensional variable *x* in terms of the new variable *y* (with support [*a*,*c*] or (*a*,*c*)) and the parameters *a* and *c*:

$$y = x(c-a) + a, \text{ therefore } \quad x = \frac{y-a}{c-a}.$$

The probability density function of the four parameter beta distribution is equal to the two parameter distribution, scaled by the range (*c-a*), (so that the total area under the density curve equals a probability of one), and with the "y" variable shifted and scaled as follows:

$$f(y;\alpha,\beta,a,c)=\frac{f(x;\alpha,\beta)}{c-a}=\frac{\left(\frac{y-a}{c-a}\right)^{\alpha-1}\left(\frac{c-y}{c-a}\right)^{\beta-1}}{(c-a)B(\alpha,\beta)}=\frac{(y-a)^{\alpha-1}(c-y)^{\beta-1}}{(c-a)^{\alpha+\beta-1}B(\alpha,\beta)}.$$

That a random variable *Y* is Beta-distributed with four parameters α, β, *a*, and *c* will be denoted by:

$$Y \sim \text{Beta}(\alpha,\beta,a,c).$$

The measures of central location are scaled (by (*c-a*)) and shifted (by *a*), as follows:

$$\text{mean}(Y) = \text{mean}(X)(c-\alpha)+\alpha=(\frac{\alpha}{\alpha+\beta})(c-a)+\alpha=\frac{\alpha c+\beta\alpha}{\alpha+\beta}$$

$$\text{mode}(Y) = \text{mode}(X)(c-\alpha)+\alpha=(\frac{\alpha-1}{\alpha+\beta-2})(c-\alpha)+\alpha=\frac{(\alpha-1)c+(\beta-1)\alpha}{\alpha+\beta-2}, \quad \text{if } \alpha,\beta>1$$

$$\text{median}(Y) = \text{median}(X)(c-\alpha)+\alpha=(I_{\frac{1}{2}}^{[-1]}(\alpha,\beta))(c-\alpha)+\alpha$$

$$\text{Wrong!!: } G_Y = G_X(c-\alpha)+\alpha=(e^{\psi(\alpha)-\psi(\alpha+\beta)})(c-\alpha)+\alpha$$

$$\text{Wrong!!: } H_Y = H_X(c-a)+\alpha=(\frac{\alpha-1}{\alpha+\beta-1})(c-\alpha)+\alpha, \quad \text{if } \alpha,\beta>0$$

(the geometric mean and harmonic mean cannot be transformed by a linear transformation in the way that the mean, median and mode can).

The statistical dispersion measures are scaled (they do not need to be shifted because they are already centered on the mean) by the range (c-a), linearly for the mean deviation and nonlinearly for the variance:

$$(\text{mean deviation around mean})(Y)=((\text{mean deviation around mean})(X))(c-a)$$

$$=\frac{2\alpha^{\alpha}\beta^{\beta}}{B(\alpha,\beta)(\alpha+\beta)^{\alpha+\beta+1}}(c-a) \quad \text{var}(Y)=\text{var}(X)(c-a)^2=\frac{\alpha\beta(c-a)^2}{(\alpha+\beta)^2(\alpha+\beta+1)}.$$

Since the skewness and excess kurtosis are non-dimensional quantities (as moments centered on the mean and normalized by the standard deviation), they are independent of the parameters *a* and *c*, and therefore equal to the expressions given above in terms of *X* (with support [0,1] or (0,1)):

$$\text{skewness}(Y)=\text{skewness}(X)=\frac{2(\beta-\alpha)\sqrt{\alpha+\beta+1}}{(\alpha+\beta+2)\sqrt{\alpha\beta}}.$$

$$\text{kurtosis excess}(Y)=\text{kurtosis excess}(X)=\frac{6[(\alpha-\beta)^2(\alpha+\beta+1)-\alpha\beta(\alpha+\beta+2)]}{\alpha\beta(\alpha+\beta+2)(\alpha+\beta+3)}$$

Kumaraswamy Distribution

In probability and statistics, the Kumaraswamy's double bounded distribution is a family of continuous probability distributions defined on the interval (0,1). It is similar to the Beta distribution, but much simpler to use especially in simulation studies since its probability density function, cumulative distribution function and quantile functions can be expressed in closed form. This distribution was originally proposed by Poondi Kumaraswamy for variables that are lower and upper bounded with a zero-inflation. This was extended to inflations at both extremes [0,1] in .

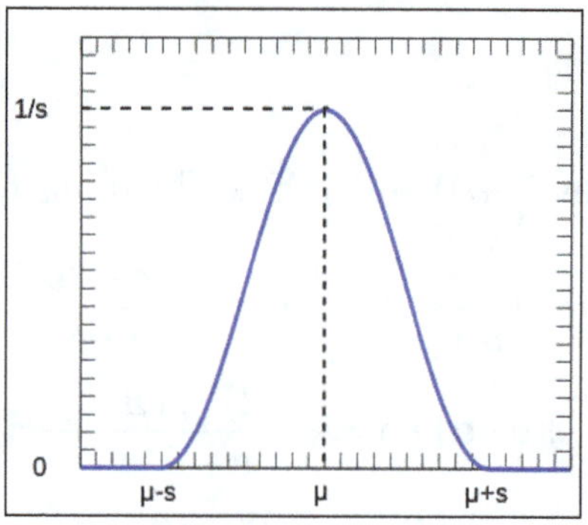

Probability density function.

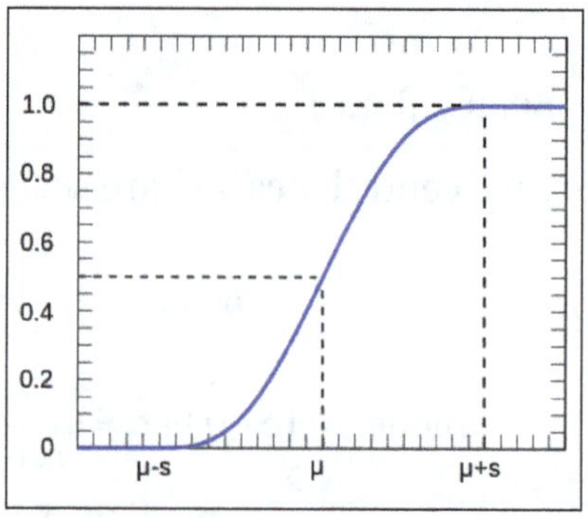

Cumulative distribution function.

Parameters	$a > 0$ (real) $b > 0$ (real)
Support	$x \in (0,1)$
PDF	$abx^{a-1}(1-x^a)^{b-1}$
CDF	$1-(1-x^a)^b$
Mean	$\frac{b\Gamma(1+\frac{1}{a})\Gamma(b)}{\Gamma(1+\frac{1}{a}+b)}$
Median	$\left(1-2^{-1/b}\right)^{1/a}$
Mode	$\left(\frac{a-1}{ab-1}\right)^{1/a}$ for $a \geq 1, b \geq 1, (a,b) \neq (1,1)$
Variance	(complicated
Skewness	(complicated)
Ex. kurtosis	(complicated)
Entropy	$(1-\frac{1}{b})+(1-\frac{1}{a})H_b - \ln(ab)$

Characterization

The probability density function of the Kumaraswamy distribution without considering any inflation is,

$$f(x;a,b) = abx^{a-1}(1-x^a)^{b-1}, \text{ where } x \in (0,1),$$

and where a and b are non-negative shape parameters.

The cumulative distribution function is,

$$F(x;a,b) = \int_0^x f(\xi;a,b)d\xi = 1-(1-x^a)^b.$$

Generalizing to Arbitrary Interval Support

In its simplest form, the distribution has a support of (0,1). In a more general form, the normalized variable x is replaced with the unshifted and unscaled variable z where:

$$x = \frac{z - z_{min}}{z_{max} - z_{min}}, \qquad z_{min} \le z \le z_{max}.$$

Properties

The raw moments of the Kumaraswamy distribution are given by:

$$m_n = \frac{b\Gamma(1+n/a)\Gamma(b)}{\Gamma(1+b+n/a)} = bB(1+n/a,b)$$

where B is the Beta function and $\Gamma(.)$ denotes the Gamma function. The variance, skewness, and excess kurtosis can be calculated from these raw moments. For example, the variance is:

$$\sigma^2 = m_2 - m_1^2.$$

The Shannon entropy (in nats) of the distribution is:

$$H = \left(1-\tfrac{1}{b}\right) + \left(1-\tfrac{1}{a}\right)H_b - \ln(ab)$$

where H_i is the harmonic number function.

Relation to the Beta Distribution

The Kumaraswamy distribution is closely related to Beta distribution. Assume that $X_{a,b}$ is a Kumaraswamy distributed random variable with parameters a and b. Then $X_{a,b}$ is the a-th root of a suitably defined Beta distributed random variable. More formally, Let

$Y_{1,b}$ denote a Beta distributed random variable with parameters $\alpha = 1$ and $\beta = b$. One has the following relation between $X_{a,b}$ and $Y_{1,b}$.

$$X_{a,b} = Y_{1,b}^{1/a},$$

with equality in distribution.

$$P\{X_{a,b} \le x\} = \int_0^x abt^{a-1}(1-t^a)^{b-1}dt = \int_0^{x^a} b(1-t)^{b-1}dt = P\{Y_{1,b} \le x^a\} = P\{Y_{1,b}^{1/a} \le x\}.$$

One may introduce generalised Kumaraswamy distributions by considering random variables of the form $Y_{\alpha,\beta}^{1/\gamma}$, with $\gamma > 0$ and where $Y_{\alpha,\beta}$ denotes a Beta distributed random variable with parameters α and β. The raw moments of this generalized Kumaraswamy distribution are given by:

$$m_n = \frac{\Gamma(\alpha+\beta)\Gamma(\alpha+n/\gamma)}{\Gamma(\alpha)\Gamma(\alpha+\beta+n/\gamma)}.$$

Note that we can re-obtain the original moments setting $\alpha = 1$, $\beta = b$ and $\gamma = a$. However, in general, the cumulative distribution function does not have a closed form solution.

Raised Cosine Distribution

In probability theory and statistics, the raised cosine distribution is a continuous probability distribution supported on the interval $[\mu - s, \mu + s]$. The probability density function (PDF) is,

$$f(x;\mu,s) = \frac{1}{2s}\left[1+\cos\left(\frac{x-\mu}{s}\pi\right)\right] = \frac{1}{s}\mathrm{hvc}\left(\frac{x-\mu}{s}\pi\right)$$

for $\mu - s \le x \le \mu + s$ and zero otherwise. The cumulative distribution function (CDF) is,

$$F(x;\mu,s) = \frac{1}{2}\left[1+\frac{x-\mu}{s}+\frac{1}{\pi}\sin\left(\frac{x-\mu}{s}\pi\right)\right]$$

for $\mu - s \le x \le \mu + s$ and zero for $x < \mu - s$ and unity for $x > \mu + s$.

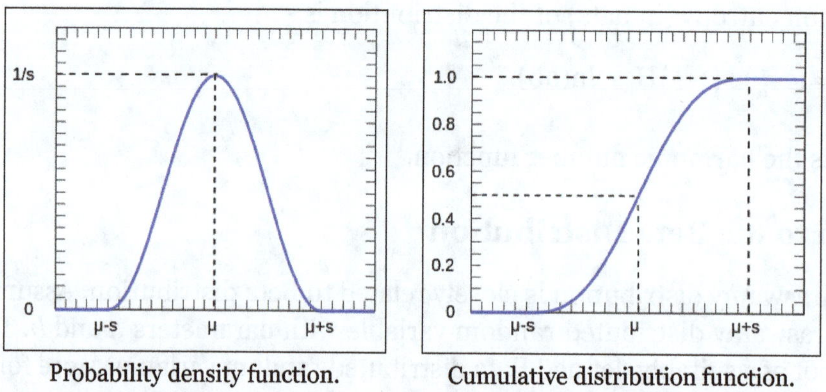

Probability density function. Cumulative distribution function.

Raised cosine	
Parameters	μ (real) $s > 0$ (real)
Support	$x \in [\mu - s, \mu + s]$
PDF	$\frac{1}{2s}\left[1+\cos\left(\frac{x-\mu}{s}\pi\right)\right] = \frac{1}{s}\mathrm{hvc}\left(\frac{x-\mu}{s}\pi\right)$
CDF	$\frac{1}{2}\left[1+\frac{x-\mu}{s}+\frac{1}{\pi}\sin\left(\frac{x-\mu}{s}\pi\right)\right]$
Mean	μ
Median	μ
Mode	
Variance	$s^2\left(\frac{1}{3}-\frac{2}{\pi^2}\right)$
Skewness	0
Ex. kurtosis	$\frac{6(90-\pi^4)}{5(\pi^2-6)^2}$
MGF	$\frac{\pi^2 \sinh(st)}{st(\pi^2+s^2t^2)}e^{\mu t}$
CF	$\frac{\pi^2 \sin(st)}{st(\pi^2-s^2t^2)}e^{i\mu t}$

The moments of the raised cosine distribution are somewhat complicated in the general case, but are considerably simplified for the standard raised cosine distribution. The standard raised cosine distribution is just the raised cosine distribution with $x > \mu + s$ and $s = 1$. Because the standard raised cosine distribution is an even function, the odd moments are zero. The even moments are given by:

$$E\left(x^{2n}\right) = \frac{1}{2}\int_{-1}^{1}\left[1+\cos(x\pi)\right]x^{2n}dx = \int_{-1}^{1} x^{2n}\mathrm{hvc}(x\pi)$$

$$= \frac{1}{n+1} + \frac{1}{1+2n}\,{}_1F_2\left(n+\frac{1}{2};\frac{1}{2},n+\frac{3}{2};\frac{-\pi^2}{4}\right)$$

where ${}_1F_2$ is a generalized hypergeometric function.

PERT Distribution

The PERT distribution (also known as the Beta-PERT distribution) gets its name because it uses the same assumption about the mean as PERT networks (used in the past for project planning). It is a version of the Beta distribution and requires the same three parameters as the Triangle distribution, namely minimum (a), mode (b) and maximum (c). The figure below shows three PERT distributions whose shape can be compared to triangle distributions:

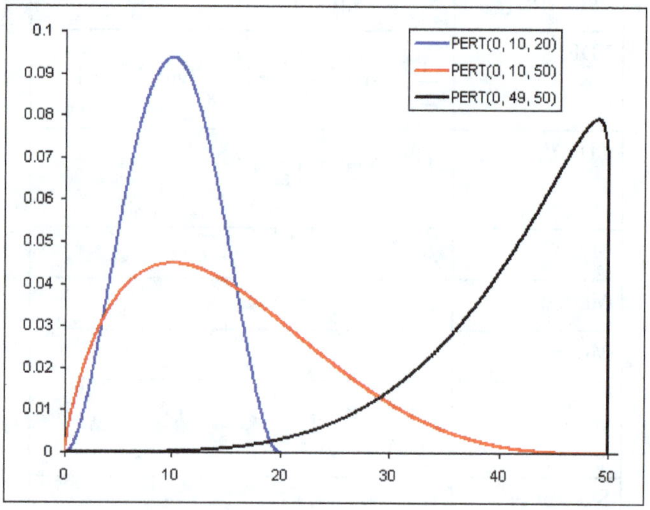

Uses

The PERT distribution is used exclusively for modeling expert estimates, where one is given the expert's minimum, most likely and maximum guesses. It is a direct alternative to a Triangle distribution, so a discussion is warranted on comparing the two:

Comparison with the Triangle Distribution

The equation of a PERT distribution is related to the Beta4 distribution as follows:

$$\text{PERT}(a, b, c) = \text{Beta4}(a1, a2, a, c)$$

where:

$$\alpha_1 = \frac{(\mu - a)*(2b - a - c)}{(b - \mu)*(c - a)}$$

$$\alpha_2 = \frac{\alpha\ 1*(c - \mu)}{(\mu - a)}$$

$$\text{The mean}(\mu) = \frac{\alpha + 4*b + c}{6}$$

The last equation for the mean is a restriction that is assumed in order to be able to determine values for a1and a2. It also shows how the mean for the PERT distribution is four times more sensitive to the most likely value than to the minimum and maximum values.

This should be compared with the Triangle distribution where the mean is equally sensitive to each parameter. The PERT distribution therefore does not suffer to the same extent the potential systematic bias problems of the Triangle distribution, that is in producing too great a value for the mean of the risk analysis results where the maximum for the distribution is very large.

The standard deviation of a PERT distribution is also less sensitive to the estimate of the extremes. Although the equation for the PERT standard deviation is rather complex, the point can be illustrated very well graphically. The figure below compares the standard deviations of the Triangle and PERT distributions with minimum a=0, maximum c= 1, and varying most likely value b.

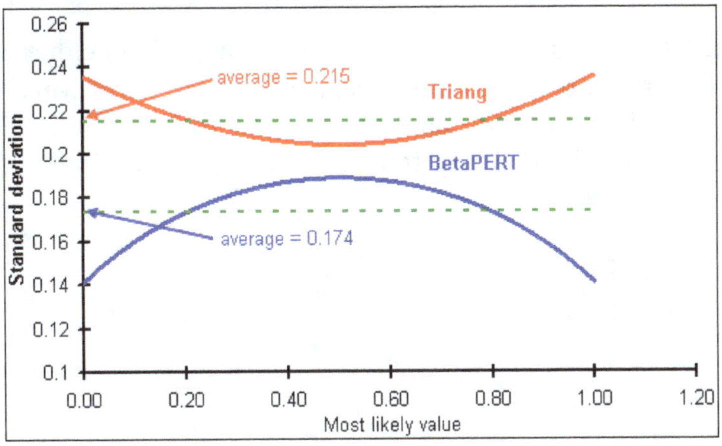

The observed pattern extends to any {a,b,c} set of values. The graph shows that the PERT distribution produces a systematically lower standard deviation than the Triangle distribution, particularly where the distribution is highly skewed (i.e. b is close to the minimum or maximum). As a general rough rule of thumb, cost and duration distributions for project tasks often have a ratio of about 2:1 between the (maximum - most likely) and (most likely - minimum), equivalent to b = 0.3333 on the figure above. The standard deviation of the PERT distribution at this point is about 88% of that for the Triangle distribution. This implies that using PERT distributions throughout a cost or schedule model, or any other additive model with similar ratios, will display about 10% less uncertainty than the equivalent model using Triangle distributions.

You might argue that the increased uncertainty that occurs with Triangle distributions will compensate to some degree for the over-confidence that is often apparent in subjective estimating. The argument is quite appealing at first sight but is not conducive to the long term improvement of the organization's ability to estimate.

Limitations to using the PERT Distribution

The PERT distribution came out of the need to describe the uncertainty in tasks during the development of the Polaris missile. The project had thousands of tasks and estimates needed to be made that were intuitive, quick and consistent in approach. The Four-Parameter Beta distribution (Beta4) was used just because it came to the author's mind (the Kumaraswamy distribution would also have been a good candidate, for example). The decision to constrain the distribution so that it's Mean = (Min + 4* Mode + Max)/6 was an approximation to their decision that the distribution should have a standard deviation of 1/6 of its range (i.e. Max - Min).

Farnum and Stanton demonstrated that, if one wishes to maintain this [standard deviation = range/6] idea then the PERT distribution should only be used with a certain range of values for the mode, namely:

$$\text{Mode} + 0.13(\text{Max} - \text{Mode}) < \text{Mode} < \text{Mode} + 0.13(\text{Max} - \text{Mode})$$

i.e. that the mode should not lie less that 13% of the range from either the Min or Max values. In practice this is pretty good advice, and tends to occur when one has a very high Max value relative to the Min and Mode, since the distribution is very skewed and gives very small density in the extreme tail making the Max value estimate rather meaningless, for example:

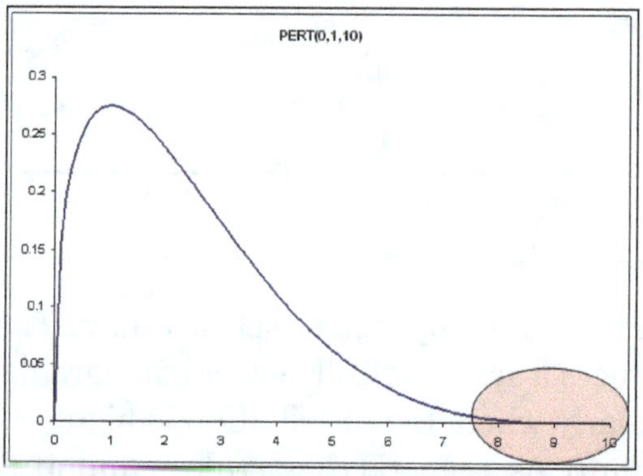

Golenko-Ginzburg describes a study that analyzed many PERT networks and concluded that "the "most likely" activity-time estimate m [mode] is practically useless". They found that the location of the mode in most project tasks was approximately one third of the distance from the Min to the Max, i.e:

$$\text{Mode} = \text{Min} + (\text{Max} - \text{Min})/3$$

Taking the Beta4(a1 ,a2,min, max) distribution again, this equates to a1 = 2, a2= 3. Thus, from Golenko-Ginzburg's viewpoint it is sufficient to use,

$$\text{Beta4}(2, 3, \text{min}, \text{max})$$

in place of,

$$\text{PERT}(\text{min}, \text{mode}, \text{max})$$

with the added advantage that one is only asking a subject matter expert for two values.

The Modified PERT Distribution

The PERT distribution, just like the Triangle, will produce just one shape from its three parameters. Thus, we are restricted to accepting this interpretation, or creating our own. The modified-PERT distribution is a quick alternative approach first proposed in Vose that modifies the weighting factor for the most likely value from 4 in a PERT to a user-defined value g:

$$\mu = \frac{a+4b+c}{6} \text{ becomes } \frac{a+yb+c}{y+2}$$

The Modified PERT distribution is widely recognized as a very useful improvement over the Triangle and PERT and is offered by many simulation packages.

References

- Park, Kun Il (2018). Fundamentals of Probability and Stochastic Processes with Applications to Communications. Springer. ISBN 978-3-319-68074-3
- Michalowicz, Joseph Victor; Nichols, Jonathan M.; Bucholtz, Frank (2013). Handbook of Differential Entropy. Chapman and Hall/CRC. p. 100. ISBN 9781466583177
- Random-variable-and-its-probability-distribution, probability, maths: toppr.com, Retrieved 11 April, 2020
- Razzaghi, Mehdi. “On the estimation of binomial success probability with zero occurrence in sample.” Journal of Modern Applied Statistical Methods 1.2 (2002): 41. url
- Fog, A. (2008). “Calculation Methods for Wallenius’ Noncentral Hypergeometric Distribution”. Communications in Statictics, Simulation and Computation. 37 (2): 258–273. doi:10.1080/03610910701790269
- Continuous-probability-distributions: sites.nicholas.duke.edu, Retrieved 19 June, 2020
- “Library and Archive catalogue”. Sackler Digital Archive. Royal Society. Archived from the original on 2011-10-25. Retrieved 2011-07-01

Stochastic Processes 5

- **Random Walk**
- **Poisson Point Process**
- **Gamma Process**
- **Branching Process**
- **Markov Processes**

A collection of random variables that are defined on a common probability space is called a stochastic process. It involves Poisson point process, gamma process, branching process, Galton-Watson process, Markov process, etc. This chapter sheds light on different stochastic processes for a thorough understanding of the subject.

A stochastic process is a family of random variables $\{x_\theta\}$, indexed by a parameter θ, where θ belongs to some index set Θ.

In almost all of the examples that we shall look at in this module, Θ will represent time. If Θ is a set of integers, representing specific time points, we have a stochastic process in discrete time and we shall replace the general subscript by n.

If Θ is the real line (or some interval of the real line) we have a stochastic process in continuous time and we shall replace the general subscript θ by t and change the notation slightly, writing X(t) rather than Xt.

The reason that we introduce the rather abstract notion of an index set Θ, rather than just working with time, is that we sometimes want to study spatial processes as well as temporal processes. In a spatial process, Θ would be a vector, representing location in space rather than time. For example, we might have a process $\left\{X_{(u,v)}\right\}$, representing a random variable that varies across two-dimensional space.

Here, $X_{(u,v)}$ represents the value of the process at position (u,v). We can even have processes that evolve in both time and space, so called spatio-temporal processes.

For processes in time, a less formal definition is that a stochastic process is simply a process that develops in time according to probabilistic rules. We shall be particularly concerned with stationary processes, in which the probabilistic rules do not change with time.

In general, for a discrete time process, the random variable X_n will depend on earlier values of the process, $X_{n-1}, X_{n-2},...$. Similarly, in continuous time, $X(t)$ will generally depend on values $X(u)$ for $u<t$.

Therefore, we are often interested in conditional distributions of the form,

$$\Pr(X_{tk} \mid X_{tk-1}, X_{tk-2}, ..., X_{t1})$$

for some set of times $t_k > t_{k-1} > ... > t_1$. In general, this conditional distribution will depend upon values of $X_{tk-1}, X_{tk-2}, ..., X_{t1}$. However, we shall focus particularly in this module on processes that satisfy the Markov property, which says that,

$$\Pr(X_{tk} \mid X_{tk-1}, X_{tk-2}, ..., X_{t1}) = \Pr(X_{tk} \mid X_{tk-1})$$

The Markov property is named after the Russian probabilist Andrei Andreyevich Markov. An informal mnemonic for remembering the Markov property is this. 'Given the present (X_{K-1}), the future (X_k) is independent of the past $(X_{k-2}, X_{k-3}, ..., X_1)$.' The Markov property is sometimes referred to as the 'lack of memory' property.

Stochastic processes that satisfy the Markov property are typically much simpler to analyse than general processes, and most of the processes that we shall study in this module are Markov processes. Of course, in attempting to model any real system it will be important to consider whether the Markov property is likely to hold.

Random Walk

A stochastic process of special form that can be interpreted as a model describing the movement of a particle in a certain state space under the action of some random mechanism. The state space is usually a d-dimensional Euclidean space or the integral lattice in it. The random mechanism can be of various kinds; the most common random walks are those generated by summation of independent random variables or by Markov chains. There is no precise generally-accepted definition of a random walk.

The trajectories of the simplest random walk in the case $d=1$ are described by the initial position $S_0 = 0$ and the sequence of sums,

$$S_n = X_1 + ... + X_n, \qquad n = 1, 2, ...,$$

where the X_i are independent and have a Bernoulli distribution:

$$P\{X_i = 1\} = p, \quad P\{X_i = 1\} = q = 1 - p, \quad p \in (0,1)$$

The value of S_n can be interpreted as the gain of one of two players after n games in each of which this player wins one dollar with probability p and loses it with probability $1-p$. If the game consists of tossing an unbiased coin, then it is assumed that $p = 1/2$ (a symmetric walk, see Bernoulli random walk). Under the assumption that the initial capital of the first player is b and that of the second player is a, the game will finish when the moving particle (with coordinates $S_1, S_2 \ldots$) first touches one of the levels a or $-b$. At this moment, one of the players is ruined. This is the classical ruin problem, in which the boundaries at points α and $-b$ can be regarded as absorbing.

In applications to queueing theory, the behaviour of the particle near the boundaries a and $-b$ can differ: e.g., if $\alpha = \infty$ and $b = 0$, then the position $Z_n + 1$ of the random particle at the moment $n+1$ is given by:

$$Z_{n+1} = \max\left(0, Z_n + X_{n+1}\right)$$

and the boundary at 0 can be called detaining or reflecting. There are also other possibilities for the behaviour of the particle in a neighbourhood of the boundaries.

If $a = \infty$, then one obtains the problem of a random walk with one boundary. If $a = b = \infty$, then one obtains an unrestricted random walk. Random walks are usually studied using the apparatus of discrete Markov chains and, in particular, by investigating the corresponding finite-difference equations. For example, in the ruin problem, let u_k be the probability that the first player is ruined, given that his initial capital is equal to k, $0 \le k \le +b$, where the total capital of both players is fixed and equal to $a+b$. Then, by the total probability formula (at the first jump), it follows that u_k satisfies the equation:

$$u_k = pu_{k+1} + qu_{k-1}, \qquad 0 < k < \alpha + b$$

and the boundary conditions $u_{\alpha+b} = 0, u_0 = 1$. Thus one obtains:

$$u_b = \frac{\left(\frac{q}{p}\right)^{\alpha+b} - \left(\frac{q}{p}\right)}{\left(\frac{q}{p}\right)^{\alpha+b} - 1} \quad \text{when } p \ne q$$

$$u_b = \frac{a}{a+b} \quad \text{when } p = q = \frac{1}{2}$$

The second of these formulas shows that even a "fair" game (in which both players have identical chances) leads to ruin with probability close to 1, provided that the capital a of the second player is large in comparison to b ($u_b = 1$ when $b < \infty$, $a < \infty$).

The ruin problem has been thoroughly investigated. For example, It was shown by J. Lagrange that the probability u_b,n of ruin of the first player at the n-th step, given that this initial capital is b, where the total capital is a+b (fixed), is equal to:

$$u_{b,n} = (a+b)^{-1} 2^n p^{(n-b)/2} q^{(n+b)/2}$$

$$\times \sum_{k=1}^{\alpha+b-1} \cos^{n-1} \frac{\pi k}{a+b} \sin \frac{\pi k}{a+b} \sin \frac{\pi bk}{a+b}.$$

The mean time before ruin of one of the players, m_b, is given by:

$$m_b = \frac{b}{q-p} - \frac{a+b}{q-p} \frac{1-\left(\frac{q}{p}\right)^b}{\left(\frac{q}{p}\right)^{\alpha-b}} \quad \text{if } p \neq q;$$

$$m_b = ab \quad \text{if } p = q = \frac{1}{2}$$

For random walks with one boundary $(a=\infty)$, described by $Z_{n+1} = \max(0, Z_n + X_{n+1})$, there is a stationary distribution for the random walk when $p<q$ and $n \to \infty$, coinciding with the distribution of the random variable S $\sup_{k \geq 0 S}$ and:

$$P\{S \geq k\} = \left(\frac{p}{q}\right)^k, \quad k = 0,1,....$$

The laws describing an unrestricted random walk follow from theorems about the behaviour of the sequence of partial sums S_n, $n = 1,2,...$ One of these laws confirms that for a symmetric random walk $(p=1/2)$, the particle hits (infinitely often) any fixed point a with probability 1. When $p<1/2$, the walk departs to the left with probability 1; in this case $S<\infty$ with probability 1.

For a symmetric random walk, the time K_n spent by the particle on the positive half-line (the number of positive terms in the sequence $S_1,...,S_n$) will be closer to or n than to $n/2$ with a large probability. This is seen from the so-called arcsine law, which implies that for large n, and that with probability 0.2, the particle spends at least 97.6% of the whole time on one side.

$$p\left\{\frac{k_n}{n} < x\right\} \approx \frac{2}{\pi} \arcsin \sqrt{x}$$

This implies, for example, that:

$$p\left\{\left|\frac{k_n}{n} - \frac{1}{2}\right| < \frac{1}{4}\right\} = 1 - 2P\left\{\frac{K_n}{n} < \frac{1}{4}\right\} \approx\sim 1 - \frac{4}{\pi}\frac{\pi}{6} = \frac{1}{3}$$

There are relations connecting random walks with boundaries with unrestricted random walks. For example, if one assumes that $Y(x) = \min\{k : S_k \geq x\}$, then:

$$P\{Y(x) = n\} =^x_n P\{S_n = x\}$$

In an unrestricted random walk, the position S_n of the particle for large n is described by the law of large numbers and the central limit theorem.

If the magnitudes $\pm$ of the jumps are changed to $\pm\Delta$ (for small Δ) and if one assumes that $p = (1+\alpha\Delta)/2$, then the position S_n of the particle after $n = t\Delta^{-2}$ steps will describe approximately (as $\Delta \to 0$) the behaviour at time t of the diffusion process with drift α, diffusion coefficient 1, and corresponding behaviour on the boundaries a and $-b$ (if these are specified for the random walk S_n).

The notion of a random walk as described above can be generalized in many ways. The simplest random walks in R^d for $d > 1$ are defined as follows. The particle leaves the origin and moves one step of length 1 in one of the 2d directions parallel to the coordinate axes. Thus, its possible positions are the points of R^d with integer coordinates. To define a random walk it is necessary to specify 2d probabilities corresponding to the different directions. One obtains a symmetric random walk if all these are equal to 1/2d. In the multi-dimensional case, the problem of a random walk with boundaries becomes much more complicated, since the shape of the boundaries becomes essentially more complex when $d > 1$. In the unrestricted case, Pólya's theorem holds: for a symmetric random walk when $d \leq 1$, the particle will sooner or later return to its initial position with probability 1 (once, and thus infinitely often); but when $d = 3$ this probability is approximately equal to 0.35 (and when $d > 3$ it is still smaller).

Another possible generalization of the simplest random walk consists of taking arbitrarily distributed independent random variables $X_1, X_2, \ldots$ in $S_n = X_1 + \ldots + X_n$, $n = 1, 2, \ldots,$. In this case, the basic qualitative laws for unrestricted random walks and for random walks with boundaries are preserved. For example, the particle reaches one of the boundaries a or $-b$ with probability 1. If $EX_i \leq 0$ and $a = \infty$, then the boundary $-b$ will be reached with probability 1. If the X_i are integer valued and $EX_i = 0$, then the particle will return to the initial position with probability 1. For arbitrarily distributed X_i with $EX_i = 0$, this statement only holds in case one considers return to an interval, not to a point.

The solution of problems concerned with the exit of random walks from an interval $(-b, a)$ turns out to be much more difficult in the general case. At the same time, these problems have many applications in mathematical statistics (sequential analysis), in the insurance business, in queueing theory, etc. In their investigation, a decisive role is played by various functionals of a random walk $\{S_n\}$, $n = 0, 1, \ldots$, called boundary functionals. Among these are $S = \sup_k \geq 0 S_k$, the time of the first passage through zero (in a positive direction), $Y+ = \min\{k : S_k > 0\}$, the time of the first passage through zero (in a negative direction), $Y\_ = \min\{k \geq 1 : S_k \leq 0\}$, the first positive sum $\chi + = S_{Y+}$, and the first non-positive sum $\chi - = S_{Y-}$, etc. It turns out that the distribution of the jumps X_i is connected with the distribution of these functionals by the following so-called factorization identity. Let $\phi(\lambda) = E_e^{i\lambda X_1}$ be the characteristic function of X_1. Then, when $|z| \leq 1$ and $\text{Im}\lambda = 0$,

$$1 - z\phi(\lambda) = \left[1 - E\left(e^{i\lambda\chi+}z^{Y}+;Y+<\infty\right)\right]$$
$$\times\left[1 - E\left(e^{i\lambda\chi}-z^{Y}-<\infty\right)\right]$$

This identity reveals the connection between boundary problems for random walks and those in the theory of functions of a complex variable, since the factors on the right-hand side of $1 - z\phi(\lambda) = \left[1 - E\left(e^{i\lambda\chi+}z^{Y}+;Y+<\infty\right)\right] \times \left[1 - E\left(e^{i\lambda\chi}-z^{Y}-<\infty\right)\right]$ are uniquely determined by those in the canonical factorization of the function $1 - z\phi(\lambda)$ on the axis $\text{Im}\lambda = 0$, that is, the expansion of this function as a product $1 - z\phi(\lambda) = A_{z+}(\lambda)A_{z-}(\lambda)$ for $1 - z\phi(\lambda)$, where $A_{z\pm}(\lambda)$ are analytic in the upper and lower half-planes, respectively, do not have zeros and are continuous there (including the boundary). In the identity $1 - z\phi(\lambda) = \left[1 - E\left(e^{i\lambda\chi+}z^{Y}+;Y+<\infty\right)\right] \times \left[1 - E\left(e^{i\lambda\chi}-z^{Y}-<\infty\right)\right]$, when $z = 1$ one can replace the first factor by:

$$\frac{1 - P\{Y+<\infty\}}{1 - Ee^{i\lambda S}} = 1 - E\left(e^{i\lambda\chi+};Y+<\infty\right)$$

Identity $1 - z\phi(\lambda) = \left[1 - E\left(e^{i\lambda\chi+}z^{Y}+;Y+<\infty\right)\right] \times \left[1 - E\left(e^{i\lambda\chi}-z^{Y}-<\infty\right)\right]$ is just one of a group of factorization identities connecting the distributions of various boundary functionals. Another one is related to the Pollaczek–Spitzer identity,

$$\sum_{n=0}^{\infty} Z^n Ee^{i\lambda S_n} = \exp\left\{\sum_{n=1}^{\infty}\frac{Z^n}{n}Ee^{i\lambda\max(0,S_n)}\right\}$$

where $\overline{S}_n = \max(0, S_1, \ldots, S_n)$. Factorization identities provide a powerful method for studying boundary problems for random walks.

Boundary problems for random walks have been investigated rather completely, including their asymptotic analysis.

Analytically, the solution of boundary problems leads to integro-difference equations. For example, the probability $u_n(x,a,b)$ that a particle starting from a point $x \in (-b,a)$ will leave this interval in time n satisfies the following equation (the total probability formula at the first jump):

$$u_{n+1}(x,a,b) = \int_{-b-x}^{a-x} u_n(x+y,a,b)\,dF(y) + 1 +$$
$$-F(a-x) + F(-b-x),$$

where $F(x) = P\{X_1 < x\}$. By passing to the generating function.

$u(z,x,a,b) = \sum_{n=1}^{\infty} z^n u_n(x,a,b)$ one obtains the usual integral equations. There are at least two approaches to the investigation of asymptotic properties of solutions of these equations. One of them is based on the study of analytic properties of the double transformation,

$$U(z,\lambda,a,b)\int e^{i\lambda x} u(z,x,a,b)dx$$

and its subsequent inversion. The other involves the methods of Vishik and Lyusternik for solving equations with a small parameter. The latter reveals profound connections between these problems and potential theory.

Much of the above carries over to the case of random walks with dependent jumps, when the random variables S_n are connected in a Markov chain, and also to multi-dimensional random walks in R^d, $d > 1$.

Poisson Point Process

In probability, statistics and related fields, a Poisson point process is a type of random mathematical object that consists of points randomly located on a mathematical space. The Poisson point process is often called simply the Poisson process, but it is also called a Poisson random measure, Poisson random point field or Poisson point field. This point process has convenient mathematical properties, which has led to it being frequently defined in Euclidean space and used as a mathematical model for seemingly random processes in numerous disciplines such as astronomy, biology, ecology, geology, seismology, physics, economics, image processing, and telecommunications.

The process is named after French mathematician Siméon Denis Poisson despite Poisson never having studied the process. Its name derives from the fact that if a collection of random points in some space forms a Poisson process, then the number of points in a region of finite size is a random variable with a Poisson distribution. The process was discovered independently and repeatedly in several settings, including experiments on radioactive decay, telephone call arrivals and insurance mathematics.

The Poisson point process is often defined on the real line, where it can be considered as a stochastic process. In this setting, it is used, for example, in queueing theory to model random events, such as the arrival of customers at a store, phone calls at an exchange or occurrence of earthquakes, distributed in time. In the plane, the point process, also known as a spatial Poisson process, can represent the locations of scattered objects such as transmitters in a wireless network, particles colliding into a detector, or trees in a forest. In this setting, the process is often used in mathematical models and in the related fields of spatial point processes, stochastic geometry, spatial statistics and continuum percolation theory. The Poisson point process can be defined on more abstract spaces. Beyond applications, the Poisson point process is an object of mathematical study in its own right. In all settings, the Poisson point process has the property that each point is stochastically independent to all the other points in the process, which is why it is sometimes called a purely or completely random process. Despite its wide use as a stochastic model of phenomena representable as points, the inherent nature of the process implies that it does not adequately describe phenomena where there is sufficiently strong interaction between the points. This has inspired the proposal of other point processes, some of which are constructed with the Poisson point process, that seek to capture such interaction.

The point process depends on a single mathematical object, which, depending on the context, may be a constant, a locally integrable function or, in more general settings, a Radon measure. In the first case, the constant, known as the rate or intensity, is the average density of the points in the Poisson process located in some region of space. The resulting point process is called a homogeneous or stationary Poisson point process. In the second case, the point process is called an inhomogeneous or nonhomogeneous Poisson point process, and the average density of points depend on the location of the underlying space of the Poisson point process. The word *point* is often omitted, but there are other *Poisson processes* of objects, which, instead of points, consist of more complicated mathematical objects such as lines and polygons, and such processes can be based on the Poisson point process.

The Poisson point process is one of the most studied and used point processes, in both the field of probability and in more applied disciplines concerning random phenomena, due to its convenient properties as a mathematical model as well as being mathematically interesting. Depending on the setting, the process has several equivalent definitions as well as definitions of varying generality owing to its many applications and characterizations.

A Poisson point process is defined on some underlying mathematical space, called a carrier space, or state space, though the latter term has a different meaning in the context of stochastic processes. The Poisson point process can be defined, studied and used in one dimension, for example, on the real line, where it can be interpreted as a counting process or part of a queueing model; in higher dimensions such as the plane where it plays a role in stochastic geometry and spatial statistics; or on more general mathematical spaces. Consequently, the notation, terminology and level of mathematical rigour used to define and study the Poisson point process and points processes in general vary according to the context. Despite all this, the Poisson point process has two key properties.

Poisson Distribution of Point Counts

A Poisson point process is characterized via the Poisson distribution. The Poisson distribution is the probability distribution of a random variable N (called a Poisson random variable) such that the probability that N equals n is given by:

$$P\{N = n\} = \frac{\Lambda^n}{n!} e^{-\Lambda}$$

where $n!$ denotes *n* factorial and the parameter Λ determines the shape of the distribution. (In fact, Λ equals the expected value of N).

By definition, a Poisson point process has the property that the number of points in a bounded region of its carrier space is a Poisson random variable.

Complete Independence

Consider a collection of disjoint and bounded subregions of the underlying space. By definition, the number of points of a Poisson point process in each bounded subregion will be completely independent of all the others.

This property is known under several names such as complete randomness, complete independence, or independent scattering and is common to all Poisson point processes. In other words, there is a lack of interaction between different regions and the points in general, which motivates the Poisson process being sometimes called a purely or completely random process.

Different Settings

In all settings where the Poisson point process is used, the key properties—the Poisson property and the independence property—play an essential role. The two properties are not logically independent; indeed, independence implies the Poisson distribution of point counts, but not the converse.

Homogeneous Poisson Point Process

If a Poisson point process has a parameter of the form $\Lambda = v\lambda$, where v is Lebesgue measure (that is, it assigns length, area, or volume to sets) and λ is a constant, then the point process is called a homogeneous or stationary Poisson point process. The parameter, called rate or intensity, is related to the expected (or average) number of Poisson points existing in some bounded region, where rate is usually used when the underlying space has one dimension. The parameter λ can be interpreted as the average number of points per some unit of extent such as length, area, volume, or time, depending on the underlying mathematical space, and it is also called the mean density or mean rate.

Interpreted as a Counting Process

The homogeneous Poisson point process, when considered on the positive half-line, can be defined as a counting process, a type of stochastic process, which can be denoted as $\{N(t), t \geq 0\}$. A counting process represents the total number of occurrences or events that have happened up to and including time t. A counting process is a homogeneous Poisson counting process with rate $\lambda > 0$ if it has the following three properties:

$N(0) = 0$; has independent increments; and the number of events (or points) in any interval of length t is a Poisson random variable with parameter (or mean) λt.

The last property implies:

$$\mathbb{E}[N(t)] = \lambda t.$$

In other words, the probability of the random variable $N(t)$ being equal to n is given by:

$$\mathbb{P}\{N(t) = n\} = \frac{(\lambda t)^n}{n!} e^{-\lambda t}.$$

The Poisson counting process can also be defined by stating that the time differences between events of the counting process are exponential variables with mean $1/\lambda$. The time differences between the events or arrivals are known as interarrival or interoccurence times.

Interpreted as a Point Process on the Real Line

Interpreted as a point process, a Poisson point process can be defined on the real line by considering the number of points of the process in the interval $(a, b]$. For the homogeneous Poisson point process on the real line with parameter $\lambda > 0$, the probability of this random number of points, written here as $N(a, b]$, being equal to some counting number n is given by:

$$P\{N(a, b] = n\} = \frac{[\lambda(b - a)]^n}{n!} e^{-\lambda(b-a)},$$

For some positive integer k, the homogeneous Poisson point process has the finite-dimensional distribution given by:

$$P\{N(a_i, b_i] = n_i, i = 1, \ldots, k\} = \prod_{i=1}^{k} \frac{[\lambda(b_i - a_i)]^{n_i}}{n_i!} e^{-\lambda(b_i - a_i)},$$

where the real numbers $a_i < b_i \leq a_{i+1}$.

In other words, $N(a, b]$ is a Poisson random variable with mean $\lambda(b - a)$, where $a \leq b$. Furthermore, the number of points in any two disjoint intervals, say, $(a_1, b_1]$ and $(a_2, b_2]$ are independent of each other, and this extends to any finite number of disjoint intervals. In the queueing theory context, one can consider a point existing (in an interval) as an *event*, but this is different to the word event in the probability theory sense. It follows that λ is the expected number of *arrivals* that occur per unit of time.

Properties

The previous definition has two important features shared by Poisson point processes in general:

- The number of arrivals in each finite interval has a Poisson distribution.
- The number of arrivals in disjoint intervals are independent random variables.

Furthermore, it has a third feature related to just the homogeneous Poisson point process:

- The Poisson distribution of the number of arrivals in each interval $(a+t,b+t]$ only depends on the interval's length $b-a$.

In other words, for any finite $t>0$, the random variable $N(a+t,b+t]$ is independent of t, so it is also called a stationary Poisson process.

Law of Large Numbers

The quantity $\lambda(b_i - a_i)$ can be interpreted as the expected or average number of points occurring in the interval $(a_i, b_i]$, namely:

$$E\{N(a_i, b_i]\} = \lambda(b_i - a_i),$$

where E denotes the expectation operator. In other words, the parameter λ of the Poisson process coincides with the *density* of points. Furthermore, the homogeneous Poisson point process adheres to its own form of the (strong) law of large numbers. More specifically, with probability one:

$$\lim_{t\to\infty} \frac{N(t)}{t} = \lambda,$$

where $\lim$ denotes the limit of a function, and λ is expected number of arrivals occurred per unit of time.

Memoryless Property

The distance between two consecutive points of a point process on the real line will be an exponential random variable with parameter λ (or equivalently, mean $1/\lambda$). This implies that the points have the memoryless property: the existence of one point existing in a finite interval does not affect the probability (distribution) of other points existing, but this property has no natural equivalence when the Poisson process is defined on a space with higher dimensions.

Orderliness and Simplicity

A point process with stationary increments is sometimes said to be orderly or regular if:

$$P\{N(t,t+\delta] > 1\} = o(\delta),$$

where little-o notation is being used. A point process is called a simple point process when the probability of any of its two points coinciding in the same position, on the underlying space, is zero. For point processes in general on the real line, the property of orderliness implies that the process is simple, which is the case for the homogeneous Poisson point process.

Martingale Characterization

On the real line, the homogeneous Poisson point process has a connection to the theory of martingales via the following characterization: a point process is the homogeneous Poisson point process if and only if is a martingale.

$$N(-\infty,t]-t,$$

Relationship to other Processes

On the real line, the Poisson process is a type of continuous-time Markov process known as a birth-death process (with just births and zero deaths) and is called a *pure* or *simple* birth process. More complicated processes with the Markov property, such as Markov arrival processes, have been defined where the Poisson process is a special case.

Restricted to the Half-line

If the homogeneous Poisson process is considered just on the half-line $[0,\infty)$, which can be the case when *t* represents time then the resulting process is not truly invariant under translation. In that case the Poisson process is no longer stationary, according to some definitions of stationarity.

Applications

There have been many applications of the homogeneous Poisson process on the real line in an attempt to model seemingly random and independent events occurring. It has a fundamental role in queueing theory, which is the probability field of developing suitable stochastic models to represent the random arrival and departure of certain phenomena. For example, customers arriving and being served or phone calls arriving at a phone exchange can be both studied with techniques from queueing theory.

Generalizations

The homogeneous Poisson process on the real line is considered one of the simplest stochastic processes for counting random numbers of points. This process can be generalized in a number of ways. One possible generalization is to extend the distribution of interarrival times from the exponential distribution to other distributions, which introduces the stochastic process known as a renewal process. Another generalization is to define the Poisson point process on higher dimensional spaces such as the plane.

Spatial Poisson Point Process

A spatial Poisson process is a Poisson point process defined in the plane $\mathbf{R}^2$. For its mathematical definition, one first considers a bounded, open or closed (or more precisely, Borel measurable) region B of the plane. The number of points of a point process N existing in this region $B \subset \mathbf{R}^2$ is a random variable, denoted by $N(B)$. If the points belong to a homogeneous Poisson process with parameter $\lambda > 0$, then the probability of n points existing in B is given by:

$$P\{N(B) = n\} = \frac{(\lambda |B|)^n}{n!} e^{-\lambda |B|}$$

where $|B|$ denotes the area of B.

For some finite integer $k \geq 1$, we can give the finite-dimensional distribution of the homogeneous Poisson point process by first considering a collection of disjoint, bounded Borel (measurable) sets $B_1, \ldots, B_k$. The number of points of the point process N existing in B_i can be written as $N(B_i)$. Then the homogeneous Poisson point process with parameter $\lambda > 0$ has the finite-dimensional distribution:

$$P\{N(B_i) = n_i, i = 1, \ldots, k\} = \prod_{i=1}^{k} \frac{(\lambda |B_i|)^{n_i}}{n_i!} e^{-\lambda |B_i|}.$$

According to one statistical study, the positions of cellular or mobile phone base stations in the Australian city Sydney, pictured above, resemble a realization of a homogeneous

Poisson point process, while in many other cities around the world they do not and other point processes are required.

The spatial Poisson point process features prominently in spatial statistics, stochastic geometry, and continuum percolation theory. This point process is applied in various physical sciences such as a model developed for alpha particles being detected. In recent years, it has been frequently used to model seemingly disordered spatial configurations of certain wireless communication networks. For example, models for cellular or mobile phone networks have been developed where it is assumed the phone network transmitters, known as base stations, are positioned according to a homogeneous Poisson point process.

Defined in Higher Dimensions

The previous homogeneous Poisson point process immediately extends to higher dimensions by replacing the notion of area with (high dimensional) volume. For some bounded region B of Euclidean space $\mathbf{R}^d$, if the points form a homogeneous Poisson process with parameter $\lambda > 0$, then the probability of n points existing in $B \subset \mathbf{R}^d$ is given by:

$$P\{N(B) = n\} = \frac{(\lambda |B|)^n}{n!} e^{-\lambda |B|}$$

where $|B|$ now denotes the d dimensional volume of B. Furthermore, for a collection of disjoint, bounded Borel sets $B_1,\ldots,B_k \subset \mathbf{R}^d$, let $N(B_i)$ denote the number of points of N existing in B_i. Then the corresponding homogeneous Poisson point process with parameter $\lambda > 0$ has the finite-dimensional distribution:

$$P\{N(B_i) = n_i, i = 1,\ldots,k\} = \prod_{i=1}^{k} \frac{(\lambda |B_i|)^{n_i}}{n_i!} e^{-\lambda |B_i|}.$$

Homogeneous Poisson point processes do not depend on the position of the underlying space through its parameter λ, which implies it is both a stationary process (invariant to translation) and an isotropic (invariant to rotation) stochastic process. Similarly to the one-dimensional case, the homogeneous point process is restricted to some bounded subset of $\mathbf{R}^d$, then depending on some definitions of stationarity, the process is no longer stationary.

Points are Uniformly Distributed

If the homogeneous point process is defined on the real line as a mathematical model for occurrences of some phenomenon, then it has the characteristic that the positions of these occurrences or events on the real line (often interpreted as time) will be uniformly distributed. More specifically, if an event occurs (according to this process) in an interval $(a,b]$ where $a \leq b$ then its location will be a uniform random variable defined on that interval. Furthermore, the homogeneous point process is sometimes called the *uniform* Poisson point process. This uniformity property extends to higher dimensions in the Cartesian coordinate, but not in, for example, polar coordinates.

Inhomogeneous Poisson Point Process

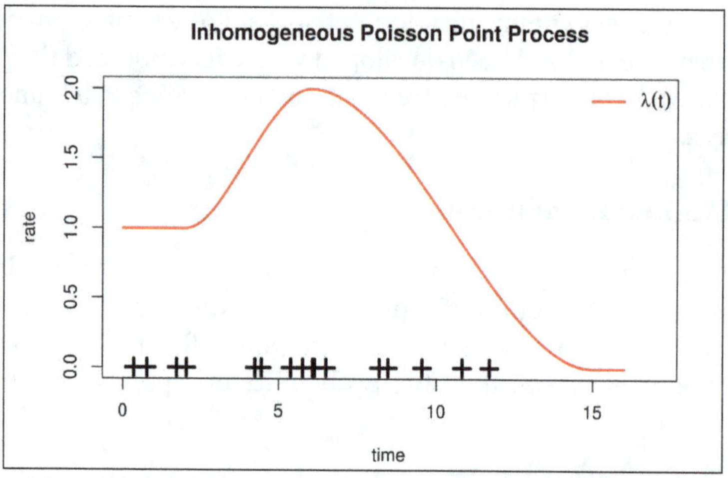

Graph of an inhomogeneous Poisson point process on the real line. The events are marked with black crosses, the time-dependent rate $\lambda(t)$ is given by the function marked red.

The inhomogeneous or nonhomogeneous Poisson point process is a Poisson point process with a Poisson parameter set as some location-dependent function in the underlying space on which the Poisson process is defined. For Euclidean space $\mathbf{R}^d$, this is achieved by introducing a locally integrable positive function $\lambda(x)$ where x is a d-dimensional point located in $\mathbf{R}^d$, such that for any bounded region B the (d-dimensional) volume integral of $\lambda(x)$ over region B is finite. In other words, if this integral, denoted by $\Lambda(B)$, is:

$$\Lambda(B) = \int_B \lambda(x)dx < \infty$$

where dx is a (d-dimensional) volume element, then for any collection of disjoint bounded Borel measurable sets $B_1,\ldots,B_k$, an inhomogeneous Poisson process with (intensity) function $\lambda(x)$ has the finite-dimensional distribution:

$$P\{N(B_i) = n_i, i = 1,\ldots,k\} = \prod_{i=1}^{k} \frac{(\Lambda(B_i))^{n_i}}{n_i!} e^{-\Lambda(B_i)}.$$

Furthermore, $\Lambda(B)$ has the interpretation of being the expected number of points of the Poisson process located in the bounded region B, namely,

$$\Lambda(B) = E[N(B)].$$

Defined on the Real Line

On the real line, the inhomogeneous or non-homogeneous Poisson point process has mean measure given by a one-dimensional integral. For two real numbers a and b, where $a \leq b$, denote by $N(a,b]$ the number points of an inhomogeneous Poisson process with intensity function $\lambda(t)$ occurring in the interval $(a,b]$. The probability of n points existing in the above interval $(a,b]$ is given by:

$$P\{N(a,b]=n\}=\frac{[\Lambda(a,b)]^n}{n!}e^{-\Lambda(a,b)}.$$

where the mean or intensity measure is:

$$(a,b)=\int_a^b \lambda(t)dt,$$

which means that the random variable $N(a,b]$ is a Poisson random variable with mean $E\{N(a,b]\}=\Lambda(a,b)$.

A feature of the one-dimension setting, is that an inhomogeneous Poisson process can be transformed into a homogeneous by a monotone transformation or mapping, which is achieved with the inverse of Λ.

Counting Process Interpretation

The inhomogeneous Poisson point process, when considered on the positive half-line, is also sometimes defined as a counting process. With this interpretation, the process, which is sometimes written as $\{N(t),t \geq 0\}$,, represents the total number of occurrences or events that have happened up to and including time t. A counting process is said to be an inhomogeneous Poisson counting process if it has the four properties:

- $N(0)=0$;
- has independent increments;
- $P\{N(t+h)-N(t)=1\}=\lambda(t)h+o(h)$;
- $P\{N(t+h)-N(t)\geq 2\}=o(h)$,

where $o(h)$ is asymptotic or little-o notation for $o(h)/h \to 0$ as $h \to 0$. In the case of point processes with refractoriness (e.g., neural spike trains) a stronger version of property applies:

$$P(N(t+h)-N(t)\geq 2)=o(h^2).$$

The above properties imply that $N(t+h)-N(t)$ is a Poisson random variable with the parameter (or mean):

$$E[N(t+h)-N(t)] = \int_{t}^{t+h} \lambda(s)ds,$$

which implies:

$$E[N(h)] = \int_{0}^{h} \lambda(s)ds.$$

Spatial Poisson Process

An inhomogeneous Poisson process defined in the plane $\mathbf{R}^2$ is called a spatial Poisson process It is defined with intensity function and its intensity measure is obtained performing an surface integral of its intensity function over some region. For example, its intensity function (as a function of Cartesian coordinates x and y) can be:

$$\lambda(x,y) = e^{-(x^2+y^2)},$$

so the corresponding intensity measure is given by the surface integral,

$$\Lambda(B) = \int_{B} e^{-(x^2+y^2)} dxdy,$$

where B is some bounded region in the plane $\mathbb{R}^2$.

In Higher Dimensions

In the plane, $\Lambda(B)$ corresponds to an surface integral while in $\mathbb{R}^d$ the integral becomes a ($d-$dimensional) volume integral.

Applications

When the real line is interpreted as time, the inhomogeneous process is used in the fields of counting processes and in queueing theory. Examples of phenomena which have been represented by or appear as an inhomogeneous Poisson point process include:

- Goals being scored in a soccer game.
- Defects in a circuit board.

In the plane, the Poisson point process is important in the related disciplines of stochastic geometry and spatial statistics. The intensity measure of this point process is dependent on the location of underlying space, which means it can be used to model phenomena with a density that varies over some region. In other words, the phenomena can be represented as points that have a location-dependent density. This processes has been used in various disciplines and uses include the study of salmon and sea lice in the oceans, forestry, and search problems.

Interpretation of the Intensity Function

The Poisson intensity function $\lambda(x)$ has an interpretation, considered intuitive, with the volume element dx in the infinitesimal sense: $\lambda(x)dx$ is the infinitesimal probability of a point of a Poisson point process existing in a region of space with volume dx located at x.

For example, given a homogeneous Poisson point process on the real line, the probability of finding a single point of the process in a small interval of width δ is approximately $\lambda\delta x$. In fact, such intuition is how the Poisson point process is sometimes introduced and its distribution derived.

Simple Point Process

If a Poisson point process has an intensity measure that is a locally finite and diffuse (or non-atomic), then it is a simple point process. For a simple point process, the probability of a point existing at a single point or location in the underlying (state) space is either zero or one. This implies that, with probability one, no two (or more) points of a Poisson point process coincide in location in the underlying space.

Simulation

Simulating a Poisson point process on a computer is usually done in a bounded region of space, known as a simulation *window*, and requires two steps: appropriately creating a random number of points and then suitably placing the points in a random manner. Both these two steps depend on the specific Poisson point process that is being simulated.

Step 1: Number of Points

The number of points N in the window, denoted here by W, needs to be simulated, which is done by using a (pseudo)-random number generating function capable of simulating Poisson random variables.

Homogeneous Case

For the homogeneous case with the constant λ, the mean of the Poisson random variable N is set to $\lambda|W|$ where $|W|$ is the length, area or (d – dimensional) volume of W.

Inhomogeneous Case

For the inhomogeneous case, $\lambda|W|$ is replaced with the (d–dimensional) volume integral,

$$\Lambda(W) = \int_W \lambda(x)dx$$

Step 2: Positioning of Points

The second stage requires randomly placing the N points in the window W.

Homogeneous Case

For the homogeneous case in one dimension, all points are uniformly and independently placed in the window or interval W. For higher dimensions in a Cartesian coordinate system, each coordinate is uniformly and independently placed in the window W. If the window is not a subspace of Cartesian space (for example, inside a unit sphere or on the surface of a unit sphere), then the points will not be uniformly placed in W, and suitable change of coordinates (from Cartesian) are needed.

Inhomogeneous Case

For the inhomogeneous, a couple of different methods can be used depending on the nature of the intensity function $\lambda(x)$. If the intensity function is sufficiently simple, then independent and random non-uniform (Cartesian or other) coordinates of the points can be generated. For example, simulating a Poisson point process on a circular window can be done for an isotropic intensity function (in polar coordinates r and θ), implying it is rotationally variant or independent of θ but dependent on r, by a change of variable in r if the intensity function is sufficiently simple.

For more complicated intensity functions, one can use an acceptance-rejection method, which consists of using (or 'accepting') only certain random points and not using (or 'rejecting') the other points, based on the ratio:

$$\frac{\lambda(x_i)}{\Lambda(W)} = \frac{\lambda(x_i)}{\int_W \lambda(x)dx.}$$

where x_i is the point under consideration for acceptance or rejection.

General Poisson Point Process

The Poisson point process can be further generalized to what is sometimes known as the general Poisson point process or general Poisson process by using a Radon measure

Λ, which is locally-finite measure. In general, this Radon measure Λ can be atomic, which means multiple points of the Poisson point process can exist in the same location of the underlying space. In this situation, the number of points at x is a Poisson random variable with mean $\Lambda(x)$. But sometimes the converse is assumed, so the Radon measure Λ is diffuse or non-atomic.

A point process N is a general Poisson point process with intensity Λ if it has the two following properties:

- The number of points in a bounded Borel set B is a Poisson random variable with mean $\Lambda(B)$. In other words, denote the total number of points located in B by $N(B)$, then the probability of random variable $N(B)$ being equal to n is given by:

$$P\{N(B)=n\}=\frac{(\Lambda(B))^n}{n!}e^{-\Lambda(B)}$$

- The number of points in n disjoint Borel sets forms n independent random variables.

The Radon measure Λ maintains its previous interpretation of being the expected number of points of N located in the bounded region B, namely,

$$\Lambda(B)=E[N(B)].$$

Furthermore, if Λ is absolutely continuous such that it has a density (which is the Radon–Nikodym density or derivative) with respect to the Lebesgue measure, then for all Borel sets B it can be written as:

$$\Lambda(B)=\int_B \lambda(x)dx,$$

where the density $\lambda(x)$ is known, among other terms, as the intensity function.

The notation of the Poisson point process depends on its setting and the field it is be ing applied in. For example, on the real line, the Poisson process, both homogeneous or inhomogeneous, is sometimes interpreted as a counting process, and the notation $\{N(t), t \geq 0\}$ is used to represent the Poisson process.

Another reason for varying notation is due to the theory of point processes, which has a couple of mathematical interpretations. For example, a simple Poisson point process may be considered as a random set, which suggests the notation $x \in N$, implying that x is a random point belonging to or being an element of the Poisson point process N. Another, more general, interpretation is to consider a Poisson or any other point process as a random counting measure, so one can write the number of points of a Poisson point process N being found or located in some (Borel measurable) region B as $N(B)$, which is a random variable. These different interpretations results in notation being used from mathematical fields such as measure theory and set theory.

For general point processes, sometimes a subscript on the point symbol, for example x, is included so one writes (with set notation) $x_i \in N$ instead of $x \in N$, and x can be used for the dummy variable in integral expressions such as Campbell's theorem, instead of denoting random points. Sometimes an uppercase letter denotes the point process, while a lowercase denotes a point from the process, so, for example, the point x or x_i belongs to or is a point of the point process X, and be written with set notation as $x \in X$ or $x_i \in X$..

Furthermore, the set theory and integral or measure theory notation can be used interchangeably. For example, for a point process N defined on the Euclidean state space $\mathbb{R}^d$ and a (measurable) function f on $\mathbb{R}^d$, the expression

$$\int_{\mathbb{R}^d} f(x)dN(x) = \sum_{x_i \in N} f(x_i),$$

demonstrates two different ways to write a summation over a point process (see also Campbell's theorem (probability)). More specifically, the integral notation on the left-hand side is interpreting the point process as a random counting measure while the sum on the right-hand side suggests a random set interpretation.

Functionals and Moment Measures

In probability theory, operations are applied to random variables for different purposes. Sometimes these operations are regular expectations that produce the average or variance of a random variable. Others, such as characteristic functions (or Laplace transforms) of a random variable can be used to uniquely identify or characterize random variables and prove results like the central limit theorem. In the theory of point processes there exist analogous mathematical tools which usually exist in the forms of measures and functionals instead of moments and functions respectively.

Laplace Functionals

For a Poisson point process N with intensity measure Λ, the Laplace functional is given by:

$$L_N(f) = e^{-\int_{\mathbb{R}^d} (1-e^{-f(x)})\Lambda(dx)},$$

$$L_N(f) = e^{-\lambda \int_{\mathbb{R}^d} (1-e^{-f(x)})dx}.$$

One version of Campbell's theorem involves the Laplace functional of the Poisson point process.

Probability Generating Functionals

The probability generating function of non-negative integer-valued random variable leads to the probability generating functional being defined analogously with respect to any non-negative bounded function v on R^d such that $0 \leq v(x) \leq 1$. For a point process N the probability generating functional is defined as:

$$G(v) = E\left[\prod_{x \in N} v(x)\right]$$

where the product is performed for all the points in N. If the intensity measure Λ of N is locally finite, then the G is well-defined for any measurable function u on R^d. For a Poisson point process with intensity measure Λ the generating functional is given by:

$$G(v) = e^{-\int_{R^d}[1-v(x)]\Lambda(dx)}$$

which in the homogeneous case reduces to,

$$G(v)e^{-\lambda\int_{R^d}[1-v(x)]dx}$$

Moment Measure

For a general Poisson point process with intensity measure Λ the first moment measure is its intensity measure:

$$M^1(B) = \Lambda(B),$$

which for a homogeneous Poisson point process with constant intensity λ means:

$$M^1(B) = \lambda |B|,$$

where $|B|$ is the length, area or volume (or more generally, the Lebesgue measure) of B.

The Mecke Equation

The Mecke equation characterizes the Poisson point process. Let N_σ be the space of all σ-finite measures on some general space Q. A point process η with intensity λ on Q is a Poisson point process if and only if for all measurable functions $f: Q \times N_\sigma \to \mathbb{R}_+$ the following holds:

$$E\left[\int f(x,\eta)\eta(dx)\right] = \int E[f(x,\eta+\delta_x)]\lambda(dx)$$

Factorial Moment Measure

For a general Poisson point process with intensity measure Λ the $n-th$ factorial moment measure is given by the expression:

$$M^{(n)}(B_1 \times .... \times B_n) = \prod_{i=1}^{n} [\Lambda(B_i)]$$

where Λ is the intensity measure or first moment measure of N, which for some Borel set B is given by,

$$\Lambda(B) = M^1(B) = E[N(B)].$$

For a homogeneous Poisson point process the n-th factorial moment measure is simply:

$$M^{(n)}(B_1 \times \cdots \times B_n) = \lambda^n \prod_{i=1}^{n} |B_i|,$$

where $|B_i|$ is the length, area, or volume (or more generally, the Lebesgue measure) of B_i. Furthermore, the $n-$th factorial moment density is:

$$\mu^{(n)}(x_1,...,x_n) = \lambda^n.$$

Avoidance Function

The avoidance function or void probability v of a point process N is defined in relation to some set B, which is a subset of the underlying space R^d, as the probability of no points of N existing in B. More precisely, for a test set B, the avoidance function is given by:

$$v(B) = P(N(B) = 0).$$

For a general Poisson point process N with intensity measure Λ, its avoidance function is given by:

$$v(B) = e^{-\Lambda(B)}$$

Rényi's Theorem

Simple point processes are completely characterized by their void probabilities. In other words, complete information of a simple point process is captured entirely in its void probabilities, and two simple point processes have the same void probabilities if and if only if they are the same point processes. The case for Poisson process is sometimes known as Rényi's theorem, which is named after Alfréd Rényi who discovered the result for the case of a homogeneous point process in one-dimension.

In one form, the Rényi's theorem says for a diffuse (or non-atomic) Radon measure Λ

on R^d and a set A is a finite union of rectangles (so not Borel) that if N is a countable subset of R^d such that:

$$P(N(A)=0)=v(A)=e^{-\Lambda(A)}$$

then N is a Poisson point process with intensity measure Λ.

Point Process Operations

Mathematical operations can be performed on point processes in order to get new point processes and develop new mathematical models for the locations of certain objects. One example of an operation is known as thinning which entails deleting or removing the points of some point process according to a rule, creating a new process with the remaining points (the deleted points also form a point process).

Thinning

For the Poisson process, the independent $p(x)-$ thinning operations results in another Poisson point process. More specifically, a $p(x)$ -thinning operation applied to a Poisson point process with intensity measure Λ gives a point process of removed points that is also Poisson point process N_p with intensity measure Λ_p, which for a bounded Borel set B is given by:

$$\Lambda_p(B)=\int_B p(x)\Lambda(dx)$$

This thinning result of the Poisson point process is sometimes known as Prekopa's theorem. Furthermore, after randomly thinning a Poisson point process, the kept or remaining points also form a Poisson point process, which has the intensity measure,

$$\Lambda_p(B)=\int_B (1-p(x))\Lambda(dx).$$

The two separate Poisson point processes formed respectively from the removed and kept points are stochastically independent of each other. In other words, if a region is known to contain n kept points (from the original Poisson point process), then this will have no influence on the random number of removed points in the same region. This ability to randomly create two independent Poisson point processes from one is sometimes known as *splitting* the Poisson point process.

Superposition

If there is a countable collection of point processes $N_1, N_2 \ldots$, then their superposition, or, in set theory language, their union, which is:

$$N=\bigcup_{i=1}^{\infty} N_i,$$

also forms a point process. In other words, any points located in any of the point processes $N_1, N_2 \ldots$ will also be located in the superposition of these point processes N.

Superposition Theorem

The superposition theorem of the Poisson point process says that the superposition of independent Poisson point processes $N_1, N_2 \ldots$ with mean measures $\Lambda_1, \Lambda_2, \ldots$ will also be a Poisson point process with mean measure:

$$\Lambda = \sum_{i=1}^{\infty} \Lambda_i.$$

In other words, the union of two (or countably more) Poisson processes is another Poisson process. If a point is sampled from a countable n union of Poisson processes, then the probability that the point x belongs to the j th Poisson process N_j is given by:

$$P(x \in N_j) = \frac{\Lambda_j}{\sum_{i=1}^{n} \Lambda_i}.$$

For two homogeneous Poisson processes with intensities $\lambda_1, \lambda_2 \ldots$, the two previous expressions reduce to,

$$\lambda = \sum_{i=1}^{\infty} \lambda_i,$$

and

$$P(x \in N_j) = \frac{\lambda_j}{\sum_{i=1}^{n} \lambda_i}.$$

Clustering

The operation clustering is performed when each point x of some point process N is replaced by another (possibly different) point process. If the original process N is a Poisson point process, then the resulting process N_c is called a Poisson cluster point process.

Random Displacement

A mathematical model may require randomly moving points of a point process to other locations on the underlying mathematical space, which gives rise to a point process operation known as displacement or translation. The Poisson point process has been used to model, for example, the movement of plants between generations, owing to the displacement theorem, which loosely says that the random independent displacement of points of a Poisson point process (on the same underlying space) forms another Poisson point process.

Displacement Theorem

One version of the displacement theorem involves a Poisson point process N on R^d with intensity function $\lambda(x)$. It is then assumed the points of N are randomly displaced somewhere else in R^d so that each point's displacement is independent and that the displacement of a point formerly at x is a random vector with a probability density $\rho(x,\cdot)$. Then the new point process N_D is also a Poisson point process with intensity function,

$$\lambda_D(y) = \int_{\mathrm{R}^d} \lambda(x)\rho(x,y)dx.$$

If the Poisson process is homogeneous with $\lambda(x) = \lambda > 0$ and if $\rho(x,y)$ is a function of $y - x$, then,

$$\lambda_D(y) = \lambda.$$

In other words, after each random and independent displacement of points, the original Poisson point process still exists.

The displacement theorem can be extended such that the Poisson points are randomly displaced from one Euclidean space R^d to another Euclidean space $\mathrm{R}^{d'}$, where $d' \geq 1$ is not necessarily equal to d.

Mapping

Another property that is considered useful is the ability to map a Poisson point process from one underlying space to another space.

Mapping Theorem

If the mapping (or transformation) adheres to some conditions, then the resulting mapped (or transformed) collection of points also form a Poisson point process, and this result is sometimes referred to as the mapping theorem. The theorem involves some Poisson point process with mean measure Λ on some underlying space. If the locations of the points are mapped (that is, the point process is transformed) according to some function to another underlying space, then the resulting point process is also a Poisson point process but with a different mean measure Λ'.

More specifically, one can consider a (Borel measurable) function f that maps a point process N with intensity measure Λ from one space S, to another space T in such a manner so that the new point process N' has the intensity measure:

$$\Lambda(B)' = \Lambda(f^{-1}(B))$$

with no atoms, where B is a Borel set and f^{-1} denotes the inverse of the function f. If N is a Poisson point process, then the new process N' is also a Poisson point process with the intensity measure Λ'.

Approximations with Poisson Point Processes

The tractability of the Poisson process means that sometimes it is convenient to approximate a non-Poisson point process with a Poisson one. The overall aim is to approximate the both number of points of some point process and the location of each point by a Poisson point process. There a number of methods that can be used to justify, informally or rigorously, approximating the occurrence of random events or phenomena with suitable Poisson point processes. The more rigorous methods involve deriving upper bounds on the probability metrics between the Poisson and non-Poisson point processes, while other methods can be justified by less formal heuristics.

Clumping Heuristic

One method for approximating random events or phenomena with Poisson processes is called the clumping heuristic. The general heuristic or principle involves using the Poisson point process (or Poisson distribution) to approximate events, which are considered rare or unlikely, of some stochastic process. In some cases these rare events are close to being independent, hence a Poisson point process can be used. When the events are not independent, but tend to occur in clusters or *clumps*, then if these clumps are suitably defined such that they are approximately independent of each other, then the number of clumps occurring will be close to a Poisson random variable and the locations of the clumps will be close to a Poisson process.

Stein's Method

Stein's method is a mathematical technique originally developed for approximating random variables such as Gaussian and Poisson variables, which has also been applied to point processes. Stein's method can be used to derive upper bounds on probability metrics, which give way to quantify how different two random mathematical objects vary stochastically. Upperbounds on probability metrics such as total variation and Wasserstein distance have been derived.

Researchers have applied Stein's method to Poisson point processes in a number of ways, such as using Palm calculus. Techniques based on Stein's method have been developed to factor into the upper bounds the effects of certain point process operations such as thinning and superposition. Stein's method has also been used to derive upper bounds on metrics of Poisson and other processes such as the Cox point process, which is a Poisson process with a random intensity measure.

Convergence to a Poisson Point Process

In general, when an operation is applied to a general point process the resulting process is usually not a Poisson point process. For example, if a point process, other than a Poisson, has its points randomly and independently displaced, then the process would not necessarily be a Poisson point process. However, under certain mathematical conditions for both the original point process and the random displacement, it has been shown via limit theorems that if the points of a point process are repeatedly displaced in a random and independent manner, then the finite-distribution of the point process will converge (weakly) to that of a Poisson point process.

Similar convergence results have been developed for thinning and superposition operations that show that such repeated operations on point processes can, under certain conditions, result in the process converging to a Poisson point processes, provided a suitable rescaling of the intensity measure (otherwise values of the intensity measure of the resulting point processes would approach zero or infinity). Such convergence work is directly related to the results known as the Palm–Khinchin equations, which has its origins in the work of Conny Palm and Aleksandr Khinchin, and help explains why the Poisson process can often be used as a mathematical model of various random phenomena.

Generalizations of Poisson Point Processes

The Poisson point process can be generalized by, for example, changing its intensity measure or defining on more general mathematical spaces. These generalizations can be studied mathematically as well as used to mathematically model or represent physical phenomena.

Poisson Point Processes on more General Spaces

For mathematical models the Poisson point process is often defined in Euclidean space, but has been generalized to more abstract spaces and plays a fundamental role in the study of random measures, which requires an understanding of mathematical fields such as probability theory, measure theory and topology.

In general, the concept of distance is of practical interest for applications, while topological structure is needed for Palm distributions, meaning that point processes are usually defined on mathematical spaces with metrics. Furthermore, a realization of a point process can be considered as a counting measure, so points processes are types of random measures known as random counting measures. In this context, the Poisson and other point processes has been studied on a locally compact second countable Hausdorff space.

Cox Point Process

A Cox point process, Cox process or doubly stochastic Poisson process is a generalization of the Poisson point process by letting its intensity measure Λ to be also random and independent of the underlying Poisson process. The process is named after David Cox who introduced it in 1955, though other Poisson processes with random intensities had been independently introduced earlier by Lucien Le Cam and Maurice Quenouille. The intensity measure may be a realization of random variable or a random field. For example, if the logarithm of the intensity measure is a Gaussian random field, then the resulting process is known as a *log Gaussian Cox process*. More generally, the intensity measures is a realization of a non-negative locally finite random measure. Cox point processes exhibit a *clustering* of points, which can be shown mathematically to be larger than those of Poisson point processes. The generality and tractability of Cox processes has resulted in them being used as models in fields such as spatial statistics and wireless networks.

Marked Poisson Point Process

An illustration of a marked point process, where the unmarked point process is defined on the positive real line, which often represents time. The random marks take on values in the state space S known as the *mark space*. Any such marked point process can be interpreted as an unmarked point process on the space $[0,\infty]\times S$. The marking theorem says that if the original unmarked point process is a Poisson point process and the marks are stochastically independent, then the marked point process is also a Poisson point process on $[0,\infty]\times S$. If the Poisson point process is homogeneous, then the gaps τ_i in the diagram are drawn from an exponential distribution.

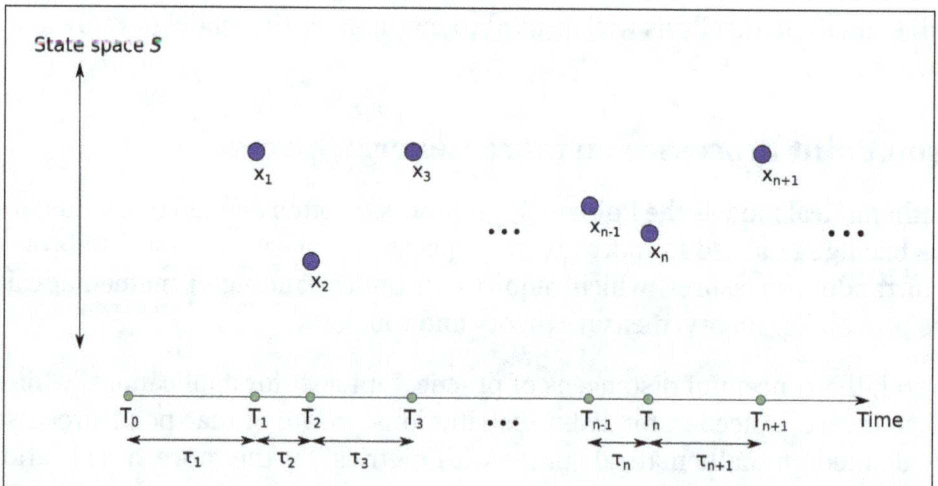

For a given point process, each random point of a point process can have a random mathematical object, known as a mark, randomly assigned to it. These marks can be as diverse as integers, real numbers, lines, geometrical objects or other point processes. The pair consisting of a point of the point process and its corresponding mark is called a marked point, and all the marked points form a marked point process. It is often assumed that the random marks are independent of each other and identically distributed, yet the mark of a point can still depend on the location of its corresponding point in the underlying (state) space. If the underlying point process is a Poisson point process, then the resulting point process is a marked Poisson point process.

Marking Theorem

If a general point process is defined on some mathematical space and the random marks are defined on another mathematical space, then the marked point process is defined on the Cartesian product of these two spaces. For a marked Poisson point process with independent and identically distributed marks, the marking theorem states that this marked point process is also a (non-marked) Poisson point process defined on the aforementioned Cartesian product of the two mathematical spaces, which is not true for general point processes.

Compound Poisson Point Process

The compound Poisson point process or compound Poisson process is formed by adding random values or weights to each point of Poisson point process defined on some underlying space, so the process is constructed from a marked Poisson point process, where the marks form a collection of independent and identically distributed non-negative random variables. In other words, for each point of the original Poisson process, there is an independent and identically distributed non-negative random variable, and then the compound Poisson process is formed from the sum of all the random variables corresponding to points of the Poisson process located in some region of the underlying mathematical space.

If there is a marked Poisson point process formed from a Poisson point process N (defined on, for example, R^d) and a collection of independent and identically distributed non-negative marks $\{M_i\}$ such that for each point x_i of the Poisson process N there is a non-negative random variable M_i, the resulting compound Poisson process is then:

$$C(B) = \sum_{i=1}^{N(B)} M_i,$$

where $B \subset \mathrm{R}^d$ is a Borel measurable set.

If general random variables $\{M_i\}$ take values in, for example, d-dimensional Euclidean space R^d, the resulting compound Poisson process is an example of a Lévy process provided that it is formed from a homogeneous Point process N defined on the non-negative numbers $[0,\infty)$.

Failure Process with the Exponential Smoothing of Intensity Functions

The failure process with the exponential smoothing of intensity functions (FP-ESI) is an extension of the nonhomogeneous Poisson process. The intensity function of an FP-ESI is an exponential smoothing function of the intensity functions at the last time points of event occurrences and outperforms other nine stochastic processes on 8 real-world failure datasets when the models are used to fit the datasets, where the model performance is measured in terms of AIC (Akaike information criterion) and BIC (Bayesian information criterion).

Gamma Process

A gamma process is a random process with independent gamma distributed increments. Often written as $\tilde{A}(t;\gamma,\lambda)$,, it is a pure-jump increasing Lévy process with intensity measure $\nu(x)=\gamma x^{-1}\exp(-\lambda x)$, for positive x. Thus jumps whose size lies in the interval $[x,x+dx)$ occur as a Poisson process with intensity $\nu(x)dx$. The parameter γ controls the rate of jump arrivals and the scaling parameter λ inversely controls the jump size. It is assumed that the process starts from a value 0 at t=0.

The gamma process is sometimes also parameterised in terms of the mean (μ) and variance (v) of the increase per unit time, which is equivalent to $\gamma=\mu^2/v$ and $\lambda=\mu/v$.

Properties

Since we use the Gamma function in these properties, we may write the process at time as $X_t \equiv \tilde{A}(t;\gamma,\lambda)$ to eliminate ambiguity.

Some basic properties of the gamma process are:

Marginal Distribution

The marginal distribution of a gamma process at time t is a gamma distribution with mean $\gamma t/\lambda$ and variance $\gamma t/\lambda^2$.

That is, its density f is given by,

$$f(x;t,\gamma,\lambda)=\frac{\lambda^{\gamma t}}{\tilde{A}(\gamma t)}x^{\gamma t-1}e^{-\lambda x}.$$

Scaling

Multiplication of a gamma process by a scalar constant α is again a gamma process with different mean increase rate.

$$\alpha\Gamma(t;\gamma,\lambda) \simeq \Gamma(t;\gamma,\lambda/\alpha)$$

Adding Independent Processes

The sum of two independent gamma processes is again a gamma process.

$$\Gamma(t;\gamma_1,\lambda) + \Gamma(t;\gamma_2,\lambda) \simeq \Gamma(t;\gamma_1+\gamma_2,\lambda)$$

Moments

$\mathbb{E}(X_t^n) = \lambda^{-n}\Gamma(\gamma t+n)/\Gamma(\gamma t), \quad n \geq 0,$ where $\Gamma(z)$ is the Gamma function.

Moment Generating Function

$$\mathbb{E}\left(\exp(\theta X_t)\right) = (1-\theta/\lambda)^{-\gamma t}, \quad \theta < \lambda$$

Correlation

$\mathrm{Corr}(X_s, X_t) = \sqrt{s/t}, s < t,$ for any gamma process $X(t)$.

The gamma process is used as the distribution for random time change in the variance gamma process.

Branching Process

In probability theory, a branching process is a type of mathematical object known as a stochastic process, which consists of collections of random variables. The random variables of a stochastic process are indexed by the natural numbers. The original purpose of branching processes was to serve as a mathematical model of a population in which each individual in generation n produces some random number of individuals in generation $n + 1$, according, in the simplest case, to a fixed probability distribution that does not vary from individual to individual. Branching processes are used to model reproduction; for example, the individuals might correspond to bacteria, each of which generates 0, 1, or 2 offspring with some probability in a single time unit. Branching processes can also be used to model other systems with similar dynamics, e.g., the spread of surnames in genealogy or the propagation of neutrons in a nuclear reactor.

A central question in the theory of branching processes is the probability of ultimate extinction, where no individuals exist after some finite number of generations. Using Wald's equation, it can be shown that starting with one individual in generation zero, the expected size of generation n equals μ^n where μ is the expected number of children of each individual. If $\mu < 1$, then the expected number of individuals goes rapidly to zero, which implies ultimate extinction with probability 1 by Markov's inequality. Alternatively, if $\mu > 1$, then the probability of ultimate extinction is less than 1 (but not necessarily zero; consider a process where each individual either has 0 or 100 children with equal probability. In that case, $\mu = 50$, but probability of ultimate extinction is greater than 0.5, since that's the probability that the first individual has 0 children). If $\mu = 1$, then ultimate extinction occurs with probability 1 unless each individual always has exactly one child.

In theoretical ecology, the parameter μ of a branching process is called the basic reproductive rate.

Mathematical Formulation

The most common formulation of a branching process is that of the Galton–Watson process. Let Z_n denote the state in period n (often interpreted as the size of generation n), and let $X_{n,i}$ be a random variable denoting the number of direct successors of member i in period n, where $X_{n,i}$ are independent and identically distributed random variables over all $n \in \{0, 1, 2,...\}$ and $i \in \{1,..., Z_n\}$. Then the recurrence equation is,

$$Z_{n+1} = \sum_{i=1}^{Z_n} X_{n,i}$$

with $Z_0 = 1$.

Alternatively, the branching process can be formulated as a random walk. Let S_i denote the state in period i, and let X_i be a random variable that is iid over all i. Then the recurrence equation is:

$$S_{i+1} = S_i + X_{i+1} - 1 = \sum_{j=1}^{i+1} X_j - i$$

with $S_0 = 1$. To gain some intuition for this formulation, imagine a walk where the goal is to visit every node, but every time a previously unvisited node is visited, additional nodes are revealed that must also be visited. Let S_i represent the number of revealed but unvisited nodes in period i, and let X_i represent the number of new nodes that are revealed when node i is visited. Then in each period, the number of revealed but unvisited nodes equals the number of such nodes in the previous period, plus the new nodes that are revealed when visiting a node, minus the node that is visited. The process ends once all revealed nodes have been visited.

Extinction Problem for a Galton Watson Process

The ultimate extinction probability is given by,

$$\lim_{n\to\infty} \Pr(Z_n = 0).$$

For any nontrivial cases (trivial cases are ones in which the probability of having no offspring is zero for every member of the population - in such cases the probability of ultimate extinction is 0), the probability of ultimate extinction equals one if $\mu \leq 1$ and strictly less than one if $\mu > 1$.

The process can be analyzed using the method of probability generating function. Let $p_0, p_1, p_2,...$ be the probabilities of producing 0, 1, 2,... offspring by each individual in each generation. Let d_m be the extinction probability by the mth generation. Obviously, $d_0 = 0$. Since the probabilities for all paths that lead to 0 by the mth generation must be added up, the extinction probability is nondecreasing in generations. That is,

$$0 = d_0 \leq d_1 \leq d_2 \leq \cdots \leq 1.$$

Therefore, d_m converges to a limit d, and d is the ultimate extinction probability. If there are j offspring in the first generation, then to die out by the mth generation, each of these lines must die out in m-1 generations. Since they proceed independently, the probability is $(d_{m-1})^j$. Thus,

$$d_m = p_0 + p_1 d_{m-1} + p_2 (d_{m-1})^2 + p_3 (d_{m-1})^3 + \cdots.$$

The right-hand side of the equation is a probability generating function. Let $h(z)$ be the ordinary generating function for p_i:

$$h(z) = p_0 + p_1 z + p_2 z^2 + \cdots.$$

Using the generating function, the previous equation becomes:

$$d_m = h(d_{m-1}).$$

Since $d_m \to d$, d can be found by solving:

$$d = h(d).$$

This is also equivalent to finding the intersection point(s) of lines $y = z$ and $y = h(z)$ for $z \geq 0$. $y = z$ is a straight line. $y = h(z)$ is an increasing (since $h'(z) = p_1 + 2p_2 z + 3p_3 z^2 + \cdots \geq 0$) and convex (since $h''(z) = 2p_2 + 6p_3 z + 12 p_4 z^2 + \cdots \geq 0$) function. There are at most two intersection points. Since (1,1) is always an intersect point for the two functions, there only exist three cases:

Three cases of $y = h(z)$ intersect with $y = z$.

Case 1 has another intersect point at $z < 1$ (see the red curve in the graph).

Case 2 has only one intersect point at $z = 1$ (See the green curve in the graph).

Case 3 has another intersect point at $z > 1$ (See the black curve in the graph).

In case 1, the ultimate extinction probability is strictly less than one. For case 2 and 3, the ultimate extinction probability equals to one.

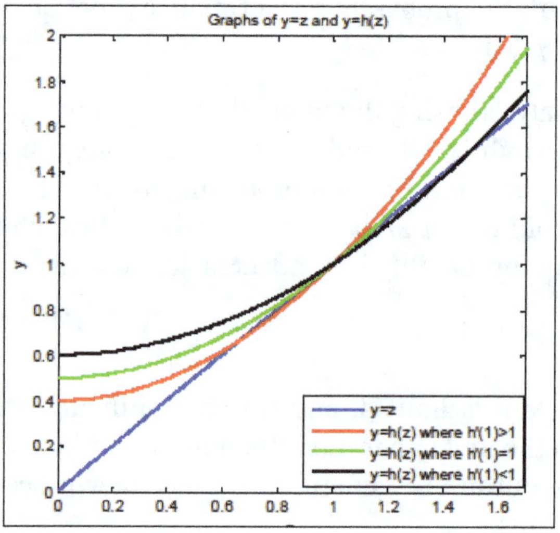

By observing that $h'(1) = p_1 + 2p_2 + 3p_3 + ... = \mu$ is exactly the expected number of offspring a parent could produce, it can be concluded that for a branching process with generating function $h(z)$ for the number of offspring of a given parent, if the mean number of offspring produced by a single parent is less than or equal to one, then the ultimate extinction probability is one. If the mean number of offspring produced by a single parent is greater than one, then the ultimate extinction probability is strictly less than one.

Size Dependent Branching Processes

Along with discussion of a more general model of branching processes known as age-dependent branching processes by Grimmett, in which individuals live for more than one generation, Krishna Athreya has identified three distinctions between size-dependent branching processes which have general application. Athreya identifies the three classes of size-dependent branching processes as sub-critical, stable, and super-critical branching measures. For Athreya, the central parameters are crucial to control if sub-critical and super-critical unstable branching is to be avoided.

Example of Extinction Problem

Consider a parent can produce at most two offspring. The extinction probability in each generation is:

$$d_m = p_0 + p_1 d_{m-1} + p_2 (d_{m-1})^2.$$

with $d_0 = 0$. For the ultimate extinction probability, we need to find d which satisfies $d = p_0 + p_1 d + p_2 d^2$.

Taking as example probabilities for the numbers of offspring produced $p_0 = 0.1$, $p_1 = 0.6$, and $p_2 = 0.3$, the extinction probability for the first 20 generations is as follows:

Generation # (1–10)	Extinction probability	Generation # (11–20)	Extinction probability
1	0.1	11	0.3156
2	0.163	12	0.3192
3	0.2058	13	0.3221
4	0.2362	14	0.3244
5	0.2584	15	0.3262
6	0.2751	16	0.3276
7	0.2878	17	0.3288
8	0.2975	18	0.3297
9	0.3051	19	0.3304
10	0.3109	20	0.331

In this example, we can solve algebraically that $d = 1/3$, and this is the value to which the extinction probability converges with increasing generations.

Simulating Branching Processes

Branching processes can be simulated for a range of problems. One specific use of simulated branching process is in the field of evolutionary biology. Phylogenetic trees, for example, can be simulated under several models, helping to develop and validate estimation methods as well as supporting hypothesis testing.

Other Branching Processes

There are many other branching processes, for example, branching processes in random environments, in which the reproduction law is chosen randomly at each generation, and multitype branching processes.

Galton–Watson Process

Galton–Watson survival probabilities for different exponential rates of population growth, if the number of children of each parent node can be assumed to follow a Poisson distribution. For $\lambda \leq 1$, eventual extinction will occur with probability 1. But the probability of survival of a new type may be quite low even if $\lambda > 1$ and the population as a whole is experiencing quite strong exponential increase.

The Galton–Watson process is a branching stochastic process arising from Francis Galton's statistical investigation of the extinction of family names. The process models family names as patrilineal (passed from father to son), while offspring are randomly either male or female, and names become extinct if the family name line dies out (holders of the family name die without male descendants). This is an accurate description of Y chromosome transmission in genetics, and the model is thus useful for understanding human Y-chromosome DNA haplogroups, and is also of use in understanding other processes; but its application to actual extinction of family names is fraught. In practice, family names change for many other reasons, and dying out of name line is only one factor, as discussed in examples, below; the Galton–Watson process is thus of limited applicability in understanding actual family name distributions.

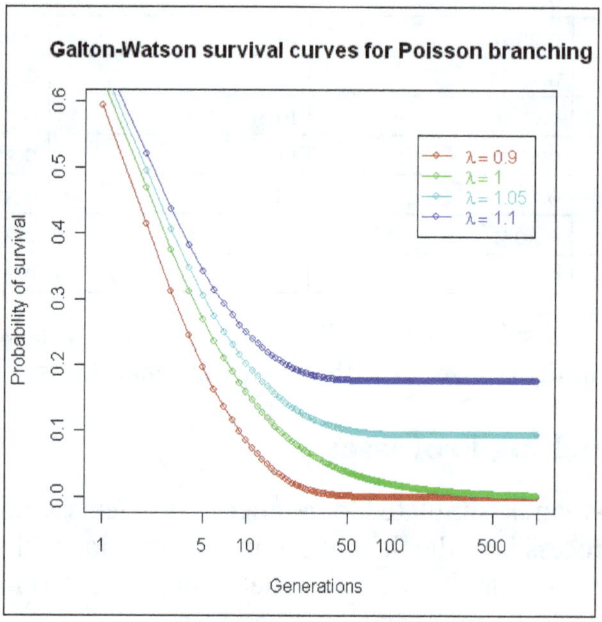

There was concern amongst the Victorians that aristocratic surnames were becoming extinct. Galton originally posed the question regarding the probability of such an event in an 1873 issue of *The Educational Times*, and the Reverend Henry William Watson replied with a solution. Together, they then wrote an 1874 paper entitled "On the probability of the extinction of families" in the Journal of the Anthropological Institute of Great Britain and Ireland (now the Journal of the Royal Anthropological Institute). Galton and Watson appear to have derived their process independently of the earlier work by I. J. Bienaymé.

Assume, for the sake of the model, that surnames are passed on to all male children by their father. Suppose the number of a man's sons to be a random variable distributed on the set { 0, 1, 2, 3,... }. Further suppose the numbers of different men's sons to be independent random variables, all having the same distribution.

Then the simplest substantial mathematical conclusion is that if the average number of a man's sons is 1 or less, then their surname will almost surely die out, and if it is more than 1, then there is more than zero probability that it will survive for any given number of generations.

Modern applications include the survival probabilities for a new mutant gene, or the initiation of a nuclear chain reaction, or the dynamics of disease outbreaks in their first generations of spread, or the chances of extinction of small population of organisms; as well as explaining (perhaps closest to Galton's original interest) why only a handful of males in the deep past of humanity now have *any* surviving male-line descendants, reflected in a rather small number of distinctive human Y-chromosome DNA haplogroups.

A corollary of high extinction probabilities is that if a lineage *has* survived, it is likely to have experienced, purely by chance, an unusually high growth rate in its early generations at least when compared to the rest of the population.

A Galton–Watson process is a stochastic process $\{X_n\}$ which evolves according to the recurrence formula $X_0 = 1$ and,

$$X_{n+1} = \sum_{j=1}^{X_n} \xi_j^{(n)}$$

where $\{\xi_j^{(n)} : n, j \in \mathbb{N}\}$ is a set of independent and identically-distributed natural number-valued random variables.

In the analogy with family names, X_n can be thought of as the number of descendants (along the male line) in the nth generation, and $\xi_j^{(n)}$ can be thought of as the number of (male) children of the jth of these descendants. The recurrence relation states that the number of descendants in the n+1st generation is the sum, over all nth generation descendants, of the number of children of that descendant.

The extinction probability (i.e. the probability of final extinction) is given by,

$$\lim_{n \to \infty} \Pr(X_n = 0).$$

This is clearly equal to zero if each member of the population has exactly one descendant. Excluding this case (usually called the trivial case) there exists a simple necessary and sufficient condition,

Extinction Criterion for Galton–Watson Process

In the non-trivial case the probability of final extinction is equal to one if $E\{\xi_1\} \leq 1$ and strictly less than one if $E\{\xi_1\} > 1$.

The process can be treated analytically using the method of probability generating functions.

If the number of children ξ_j at each node follows a Poisson distribution with parameter

λ, a particularly simple recurrence can be found for the total extinction probability x_n for a process starting with a single individual at time $n = 0$:

$$x_{n+1} = e^{\lambda(x_n - 1)},$$

giving the above curves.

Bisexual Galton–Watson Process

In the classical Galton–Watson process described above, only men are considered, effectively modeling reproduction as asexual. A model more closely following actual sexual reproduction is the so-called "bisexual Galton–Watson process", where only couples reproduce. (*Bisexual* in this context refers to the number of sexes involved, not sexual orientation.) In this process, each child is supposed as male or female, independently of each other, with a specified probability, and a so-called "mating function" determines how many couples will form in a given generation. As before, reproduction of different couples are considered to be independent of each other. Now the analogue of the trivial case corresponds to the case of each male and female reproducing in exactly one couple, having one male and one female descendant, and that the mating function takes the value of the minimum of the number of males and females (which are then the same from the next generation onwards).

Since the total reproduction within a generation depends now strongly on the mating function, there exists in general no simple necessary and sufficient condition for final extinction as is the case in the classical Galton–Watson process. However, excluding the non-trivial case, the concept of the averaged reproduction mean (Bruss (1984)) allows for a general sufficient condition for final extinction,

Extinction Criterion

If in the non-trivial case the averaged reproduction mean per couple stays bounded over all generations and will not exceed 1 for a sufficiently large population size, then the probability of final extinction is always 1.

Examples:

Citing historical examples of Galton–Watson process is complicated due to the history of family names often deviating significantly from the theoretical model. Notably, new names can be created, existing names can be changed over a person's lifetime, and people historically have often assumed names of unrelated persons, particularly nobility. Thus, a small number of family names at present is not in itself evidence for names having become extinct over time, or that they did so due to dying out of family name lines – that requires that there were more names in the past *and* that they die out due to the line dying out, rather than the name changing for other reasons, such as vassals assuming the name of their lord.

Chinese names are a well-studied example of surname extinction: there are currently only about 3,100 surnames in use in China, compared with close to 12,000 recorded in the past, with 22% of the population sharing the names Li, Wang and Zhang (numbering close to 300 million people), and the top 200 names covering 96% of the population. Names have changed or become extinct for various reasons such as people taking the names of their rulers, orthographic simplifications, taboos against using characters from an emperor's name, among others. While family name lines dying out may be a factor in the surname extinction, it is by no means the only or even a significant factor. Indeed, the most significant factor affecting the surname frequency is other ethnic groups identifying as Han and adopting Han names. Further, while new names have arisen for various reasons, this has been outweighed by old names disappearing.

By contrast, some nations have adopted family names only recently. This means both that they have not experienced surname extinction for an extended period, and that the names were adopted when the nation had a relatively large population, rather than the smaller populations of ancient times. Further, these names have often been chosen creatively and are very diverse. Examples include:

- Japanese names, which in general use date only to the Meiji restoration in the late 19th century (when the population was over 30,000,000), have over 100,000 family names, surnames are very varied, and the government restricts married couples to using the same surname.
- Many Dutch names have included a family name only since the Napoleonic Wars in the early 19th century, and there are over 68,000 Dutch family names.
- Thai names have included a family name only since 1920, and only a single family can use a given family name; hence there are a great number of Thai names. Furthermore, Thai people change their family names with some frequency, complicating the analysis.

On the other hand, some examples of high concentration of family names is not primarily due to the Galton–Watson process:

- Vietnamese names have about 100 family names, and 60% of the population sharing three family names. The name Nguyễn alone is estimated to be used by almost 40% of the Vietnamese population, and 90% share 15 names. However, as the history of the Nguyễn name makes clear, this is in no small part due to names being forced on people or adopted for reasons unrelated to genetic relation.

Markov Processes

A stochastic process whose evolution after a given time does not depend on the evolution before t, given that the value of the process at t is fixed (briefly; the "future" and "past" of the process are independent of each other for a known "present").

The defining property of a Markov process is commonly called the Markov property; it was first stated by A.A. Markov. However, in the work of L. Bachelier it is already possible to find an attempt to discuss Brownian motion as a Markov process, an attempt which received justification later in the research of N. Wiener. The basis of the general theory of continuous-time Markov processes was laid by A.N. Kolmogorov.

The Markov Property

There are essentially distinct definitions of a Markov process. One of the more widely used is the following. On a probability space (Ω, F, P) let there be given a stochastic process $X(t)$, $t \in T$, taking values in a measurable space $(E, \mathcal{B})$, where T is a subset of the real line **R**. Let N_t (respectively, N^t) be the σ-algebra in Ω generated by the variables $X(s)$ for $S \leq t$ ($S \geq t$), where $s \in T$. In other words, N_t (respectively, N^t) is the collection of events connected with the evolution of the process up to time (starting from time) t. $X(t)$ is called a Markov process if (almost certainly) for all $t \in T$, $\Lambda_1, \Lambda_2 \in N^t$ the Markov property,

$$P\{\Lambda_1, \Lambda_2 \mid X(t)\} = P\{\Lambda_1 \mid X(t)\} P\{\Lambda_2 \mid X(t)\}$$

holds, or, what is the same, if for any $t \in T$ and $\Lambda \in N^t$,

$$P\{\Lambda \mid N^t\} = \{\Lambda \mid X(t)\}$$

A Markov process for which T is contained in the natural numbers is called a Markov chain (however, the latter term is mostly associated with the case of an at most countable E). If is an interval in **R** and E is at most countable, a Markov process is called a continuous-time Markov chain. Examples of continuous-time Markov processes are furnished by diffusion processes and processes with independent increments, including Poisson and Wiener processes.

In what follows the discussion will concern only the case $T = [0, \infty)$, for the sake of being specific. The formulas $P\{\Lambda_1, \Lambda_2 \mid X(t)\} = P\{\Lambda_1 \mid X(t)\} P\{\Lambda_2 \mid X(t)\}$ and $P\{\Lambda \mid N^t\} = \{\Lambda \mid X(t)\}$ give an explicit interpretation of the principle of independence of "past" and "future" events when the "present" is known, but the definition of Markov process based on them has proved to be insufficiently flexible in the numerous situations where one is obliged to consider not one, but a collection of conditions of the type $P\{\Lambda_1, \Lambda_2 \mid X(t)\} = P\{\Lambda_1 \mid X(t)\} P\{\Lambda_2 \mid X(t)\}$ or $P\{\Lambda \mid N^t\} = \{\Lambda \mid X(t)\}$ corresponding to different, but in some sense consistent, measures P. Such reasoning has led to the acceptance of the following definitions.

Suppose one is given:

- A measurable space $(E, \mathcal{B})$, where the σ-algebra $\mathcal{B}$ contains all one-point sets in E;

- A measurable space (Ω,F), equipped with a family of σ-algebras $F_t^s \subset F, o \le s \le t\infty$, such that $F_t^s \subset F_v^u$ if $[s,t] \subset [u,v]$.
- A function ("trajectory") $x_t = x_t(\omega)$, defining for $t \in [o,\infty)$ and $v \in [o,\infty)$ a measurable mapping from (Ω, F_t^v) to $(E,\mathcal{B})$.
- For each $s \ge o$ and $x \in E$ a probability measure $P_{s,x}$ on the σ-algebra F_∞^s such that the function $P(s,;t,B) = P_s,\{x_t \in B\}$ is measurable with respect to $\mathcal{B}$, if $s \in [o,t]$ and $B \in \mathcal{B}$.

The collection $X(t) = (x_t, F_t^s, P_{s,x})$ is called a (non-terminating) Markov process given on $(E,\mathcal{B})$ if $P_{s,x}$-almost certainly,

$$P_{s,x}\{\Lambda \mid F_t^s\} = P_{t,x_t}\{\Lambda\}$$

for any $o \le s \le t$ and $\Lambda \in N^t$. Here Ω is the space of elementary events, $(E,\mathcal{B})$ is the phase space or state space and $P(s,x,t,B)$ is the transition function or transition probability of $X(t)$. If E is endowed with a topology and is the collection of Borel sets in E, then it is commonly said that the Markov process is given on E. Usually included in the definition of a Markov process is the requirement that $P(s,x;s,\{x\}) \equiv 1$, and then $P_{s,x}\{\Lambda\} \Lambda \in F_\infty^s$, $\Lambda \in F_\infty^s$, is interpreted as the probability of Λ under the condition that $x_s = x$.

The following question arises: Is every Markov transition function $P(s,x,t,B)$, given on a measurable space $(E,\mathcal{B})$, the transition function of some Markov process? The answer is affirmative if, for example, E is a separable, locally compact space and $\mathcal{B}$ is the family of Borel sets in E. In addition, let E be a complete metric space and let,

$$\lim_{h \downarrow o} \alpha_\epsilon(h) = o$$

for any $\epsilon > o$, where,

$$\alpha_\epsilon(h) = \sup\{P(s,x;t,V_\epsilon(x)) : x \in E, o < t-s < h\}$$

and $V_\epsilon(x)$ is the complement of the ϵ-neighbourhood of x. Then the corresponding Markov process can be taken to be right-continuous and having left limits (that is, its trajectories can be chosen so). The existence of a continuous Markov process is guaranteed by the condition $\alpha_\epsilon(h) = o(h)$ as $h \downarrow o$.

In the theory of Markov processes most attention is given to homogeneous (in time) processes. The corresponding definition assumes one is given a system of objects a)–d) with the difference that the parameters s and u may now only take the value o. Even the notation can be simplified:

$$P_x = P_{ox}, \quad F_t =_t^o, \quad P(t,x,B) = P(o,x;t,B)$$

$$x \in E, \quad t \ge o, \quad B \in \mathcal{B}$$

Subsequently, homogeneity of Ω is assumed. That is, it is required that for any $\omega \in \Omega$ and $s \geq 0$ there is an $\omega' \in \Omega$ such that $x_t(\omega') = x_{t+s}(\omega)$ for $t \geq 0$. Because of this, on the σ-algebra **N**, the smallest σ-algebra in Ω containing the events $\{\omega : x_s \in B\}$, the time shift operators θ_t are defined, which preserve the operations of union, intersection and difference of sets, and for which,

$$\theta_t\{\omega : x_s \in B\} = \{\omega : x_{t+s} \in B\}$$

where $s, t \geq 0 B \in \mathcal{B}$.

The collection $X(t) = (xt, F_t, P_x)$ is called a (non-terminating) homogeneous Markov process given on $(E, \mathcal{B})$ if P_x-almost certainly,

$$P_x\{\theta_t \Lambda \mid F_t\} = P_{x_t}\{\Lambda\}$$

for $x \in E$, $t \geq 0$ and $\Lambda \in N$. The transition function of $X(t)$ is taken to be $P(t, x, B)$, where, unless otherwise indicated, it is required that $P(0, x, \{x\}) \equiv 1$. It is useful to bear in mind that in the verification of $P_x\{\theta_t \Lambda \mid F_t\} = P_{x_t}\{\Lambda\}$ it is only necessary to consider sets of the form $\Lambda = \{\omega : x_s \in B\}$, where $s \geq 0$, $B \in \mathcal{B}$, and in $P_x\{\theta_t \Lambda \mid F_t\} = P_{x_t}\{\Lambda\}$, F_t may always replaced by the σ-algebra F_t equal to the intersection of the completions of F_t relative to all possible measures $P_x\{x \in B\}$. Often, one fixes on $\mathcal{B}$ a probability measure μ (the "initial distribution") and considers a random Markov function (x_t, F_t, P_μ), where P_μ is the measure on F_∞ given by:

$$P_\mu\{\cdot\} = \int P_x\{\cdot\}\mu(dx)$$

A Markov process $X(t) = (x_t, F_t, P_x)$ is called progressively measurable if for each $t > 0$ the function $x(s, \omega) = x_s(\omega)$ induces a measurable mapping from $([0, t] \times \Omega, \mathcal{B}_t \times F_t)$ to $(E, \mathcal{B})$, where $\mathcal{B}_t$ is the σ-algebra of Borel subsets of $[0, t]$. A right-continuous Markov process is progressively measurable. There is a method for reducing the non-homogeneous case to the homogeneous case, and in what follows homogeneous Markov processes will be discussed.

PERMISSIONS

All chapters in this book are published with permission under the Creative Commons Attribution Share Alike License or equivalent. Every chapter published in this book has been scrutinized by our experts. Their significance has been extensively debated. The topics covered herein carry significant information for a comprehensive understanding. They may even be implemented as practical applications or may be referred to as a beginning point for further studies.

We would like to thank the editorial team for lending their expertise to make the book truly unique. They have played a crucial role in the development of this book. Without their invaluable contributions this book wouldn't have been possible. They have made vital efforts to compile up to date information on the varied aspects of this subject to make this book a valuable addition to the collection of many professionals and students.

This book was conceptualized with the vision of imparting up-to-date and integrated information in this field. To ensure the same, a matchless editorial board was set up. Every individual on the board went through rigorous rounds of assessment to prove their worth. After which they invested a large part of their time researching and compiling the most relevant data for our readers.

The editorial board has been involved in producing this book since its inception. They have spent rigorous hours researching and exploring the diverse topics which have resulted in the successful publishing of this book. They have passed on their knowledge of decades through this book. To expedite this challenging task, the publisher supported the team at every step. A small team of assistant editors was also appointed to further simplify the editing procedure and attain best results for the readers.

Apart from the editorial board, the designing team has also invested a significant amount of their time in understanding the subject and creating the most relevant covers. They scrutinized every image to scout for the most suitable representation of the subject and create an appropriate cover for the book.

The publishing team has been an ardent support to the editorial, designing and production team. Their endless efforts to recruit the best for this project, has resulted in the accomplishment of this book. They are a veteran in the field of academics and their pool of knowledge is as vast as their experience in printing. Their expertise and guidance has proved useful at every step. Their uncompromising quality standards have made this book an exceptional effort. Their encouragement from time to time has been an inspiration for everyone.

The publisher and the editorial board hope that this book will prove to be a valuable piece of knowledge for students, practitioners and scholars across the globe.

INDEX